#1 Drone Cadets Tech Series

Soldering Made Simple

A Step-By-Step Guide to Soldering a Circuit Board

M Grace Cantwell & Robert Stipak

Drone Cadets

Published in the United States by Drone Cadets LLC, New York.
Photo Credits Eduardo Escamilla and M Grace Cantwell
Model Credit: Camila Pena and Ashley Rodriguez
Fonts: Myriad Pro
Drone Cadets is a registered Trademark of Drone Cadets, LLC.
The Drone Cadets mission is to empower underserved communities worldwide by providing accessible Drone STEM Technology to help close the diversity gap in the workforce. For more information visit DroneCadets.com

Table of Contents

About the Drone Cadets Tech Series

Drone Cadets is a unique **Drone Education Program** designed to produce safe and responsible drone pilots of any age. Our Certified Drone Instructors have taught thousands of students. We work with schools and organizations throughout the greater NY area and beyond to bring life-changing technology education to families and under-served populations throughout the world. Visit us at https://DroneCadets.com

Welcome to the Drone Cadets Tech Series! Here, you will find the classroom reference materials needed to understand the fundamentals of drone technology. Whether you are a student in our Drone Flight School program or looking to learn on your own, our materials will help you learn to fly, repair, and build drones. The Drone Cadets Tech Series will help you gain the knowledge and skills needed to enter into, and become proficient in, the world of drone technology.

Soldering Made Simple

Soldering Made Simple, A Step-By-Step Guide to Soldering a Circuit Board, is an essential workbook for those looking to learn the basics of soldering electronics. Written in an easy-to-follow format, this comprehensive workbook walks you through the entire soldering process from start to finish. It covers topics such as basic safety, soldering tools, and soldering techniques. With detailed step-by-step instructions, quizzes, and puzzles, Soldering Made Simple is an invaluable resource for anyone looking to learn the basics of soldering. Plus, the included lab activities will give you hands-on experience and build confidence in your new found soldering skills.

Chapter 1: Introduction to Soldering

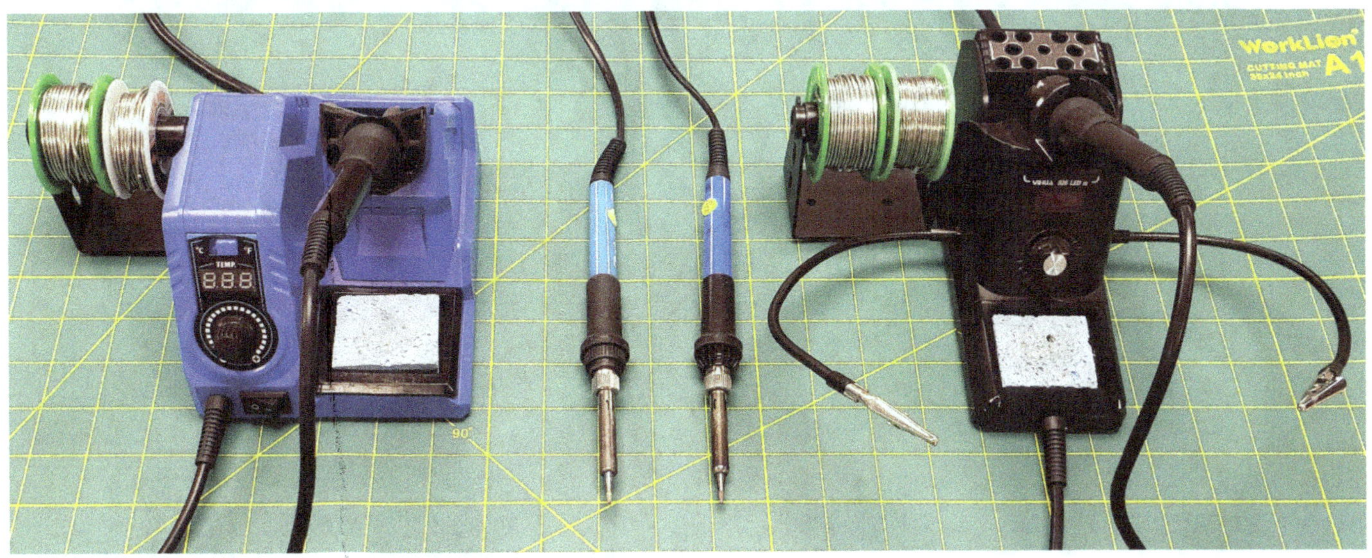

Soldering is a process used to join two or more metal components together. It is a highly-skilled technique that is used to secure electrical connections and create permanent bonds between components. Soldering is often used to assemble circuit boards, build electronic devices and repair broken electrical connections. Soldering involves heating a metal alloy, called solder, and applying it to the two pieces of metal to be joined. The solder melts and creates a bond between the two pieces of metal. It is important to use the correct type of solder and soldering techniques to ensure a strong and reliable connection.

Types of Solder

There are two primary types of solder: lead-based and tin-based. Lead-based solder is typically composed of a combination of Tin (Sn) and Lead (Pb), and is the most commonly used in general applications. Tin-based solder is composed of tin and other metals such as copper, silver, and zinc. Tin-based solder is typically used in more specialized applications, such as electronics, and is less prone to corrosion than lead-based solder.

We recommend tin-based solder over lead because it is less toxic and more environmentally friendly. Lead is a known neurotoxin, and because of its toxicity, it is being phased out of many applications. Tin-based solder is also more effective and reliable, making it the best option for most applications. If for any reason you come in contact with lead-based solder, it is important to wash your hands thoroughly and avoid touching your face. Always work in a well ventilated area and do not breathe in the smoke from any type of solder, especially lead-based solder.

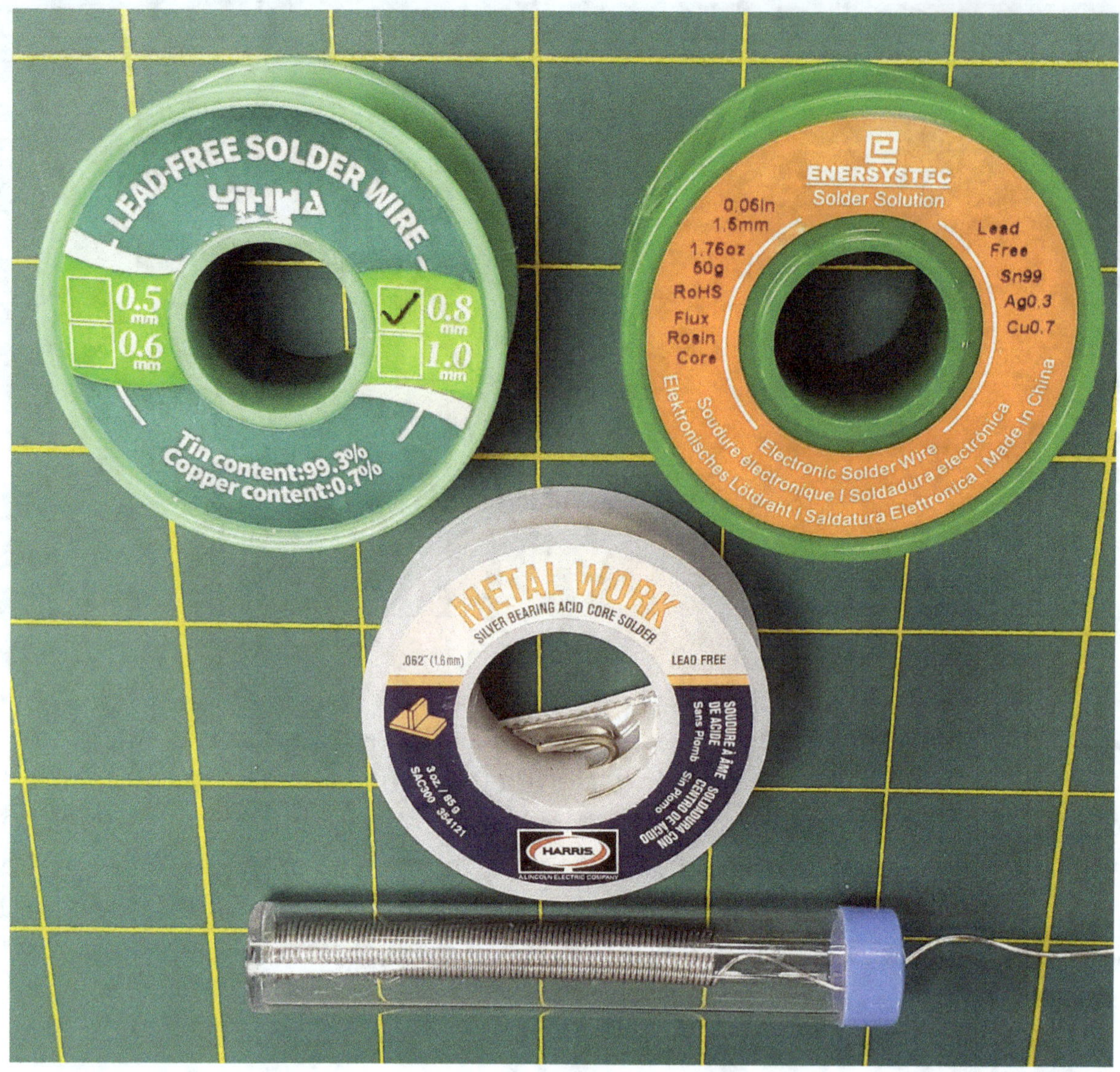

To learn to solder we will begin by practicing on a circuit board. For this we recommend Lead-Free, rosin core Solder Wire, gauge .5mm to 1.5mm.

Acid-Free Rosin Flux

Flux is a chemical agent used to help solder metals together. It helps to prevent oxidation of the metal parts during the soldering process and improves the wettability of the joint, which means it makes the solder spread over the metal more easily. The flux actually helps pull the solder onto the surface of the metal. Flux is typically in liquid or paste form and is applied to the parts prior to soldering them together.

While flux is essential for soldering, Acid-Free Rosin Flux is specifically designed for use with lead-free solders and electronic components. It is a non-corrosive, neutral flux that does not contain any acids, and is designed to prevent damage to delicate electronic components from the flux residue.

To use Rosin Flux, first clean the parts to be joined with a brush or swab and a small amount of isopropyl alcohol. Then, apply Rosin Flux to the parts. Next, heat the parts with a soldering iron, then add solder and let it melt into the joint. Finally, let the parts cool before handling.

When using Rosin-Core Solder, adding more flux may not be necessary for most jobs, although you'll still want to use alcohol to clean away the residue left behind.

Check on Learning: Solder and Flux

1.   *What are the two primary types of solder?*
   a.   *Acidic and basic*
   b.   *Soft and hard*
   c.   *Lead and tin*
   d.   *Tin and steel*

2.   *Why might you want to use one type of solder over the other?*
   a.   *One is less acidic*
   b.   *One is less expensive*
   c.   *One is more malleable*
   d.   *One is non-toxic*

3.   *What is one purpose of using flux?*
   a.   *To clean the surface of the metal*
   b.   *To prevent oxidation*
   c.   *To make the metal more malleable*
   d.   *None of the above*

4.   *Why is Acid-Free Rosin Flux recommended for electronic components?*
   a.   *It is less corrosive than other types of flux*
   b.   *It is more effective than other types of flux*
   c.   *It is cheaper than other types of flux*
   d.   *None of the above*

5.   *What is the periodic table abbreviation for lead*
   a.   *Le*
   b.   *Pb*
   c.   *Sn*
   d.   *Au*

Soldering Iron Tips

There are many different types of soldering tips available on the market. When selecting a soldering iron tip, it is important to consider the type of surface that you are working on and the size of the area to be soldered. Generally, it is best to use a smaller tip for soldering small, delicate components, and a larger tip for soldering larger, more rugged components.

The most commonly used tips include:

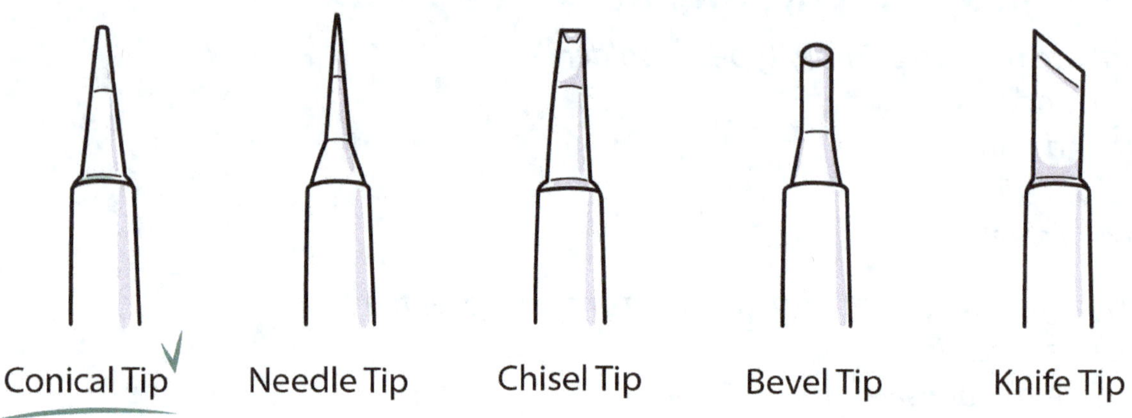

Conical Tip: This is a conical-shaped tip that is used for precise soldering of small components and fine wires. This is the recommended tip for electronics because it provides greater accuracy and control.

Needle Tip: This is a small, pointed tip that is used for intricate soldering tasks, such as wires and small soldering joints.

Chisel Tip: This is a wide, flat tip, like a flat head screwdriver, that is used for soldering large surfaces, like large components to circuit boards. It makes spreading solder easier.

Bevel Tip: This is a curved flat tip, as if you cut off the point of a conical tip, also known as a hoof tip. The bevel tip is used for soldering components with hard-to-reach areas. Its wide surface area heats up areas quickly and it's great for medium gauge wire, and desoldering, but not for small wires like we usually find on a drone motor.

Knife Tip: The knife has a thin slanted edge and a fine pointed tip, like a cutting blade. This versatile tip can be used to heat up connections or to create narrow solder lines on a circuit board. It is also useful for desoldering and removing components from the board.

How to Pick the Right Tip

When selecting the right tip for soldering you need to think about the size of the area you're working with, the size of the piece you're soldering, the surface you need to heat, the other parts and connections around it, and the amount of solder you need to use.

You want to use the biggest tip available that will fit into the space in which you are working.

For example, a Bevel or Chisel tip, with its larger surface area, will heat solder more quickly while desoldering than a Needle tip, whose tiny surface area prevents it from melting all the solder. It just takes too long for the heat to spread.

If you're not sure, the conical tip is the most versatile and popular of all the tips, as it is ideal for both precision work and larger soldering jobs. We recommended it for electronics because it provides greater accuracy and control when soldering small components. The conical shape allows the user to better focus the heat onto the desired area and prevents excess heat from damaging surrounding components.

Check on Learning: Soldering Tips

1.	*What type of soldering iron tip is most popular and can be used to solder most electronic components?*
    a.	*Flat*
    b.	*Bevel*
    c.	*Needle*
    d.	*Conical*

2.	*What should you consider when choosing a tip for an operation?*
    a.	*The area and type of surface being worked on*
    b.	*The age of the soldering iron*
    c.	*The amount of time you will be soldering*
    d.	*The size of your hands*

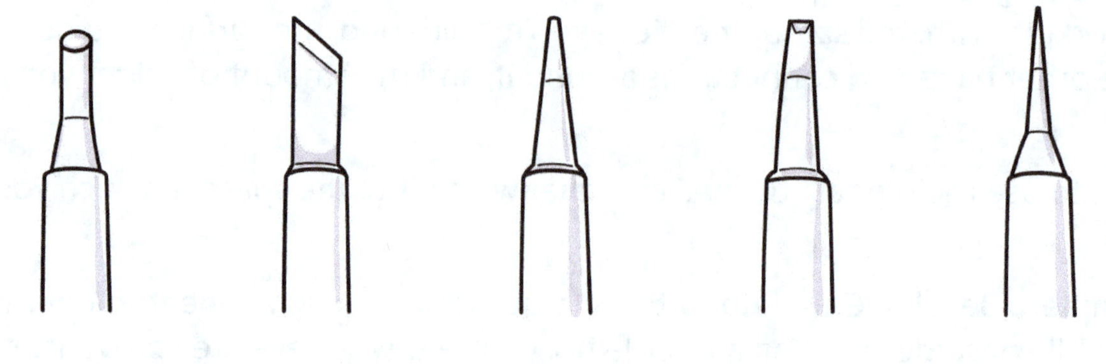

Chapter 2: Before You Begin

Gather your supplies and prepare your work area before you begin. Put on safety glasses, gloves and long sleeves. Soldering can be done safely by following the recommended warnings and procedures below. It's a great idea to keep your supplies together for future use and find an area that can be used as a soldering station.

Always Keep Safety First

At Drone Cadets, we have an oath to remind us of the safety rules, and the last line is; **"I will always keep safety first!"** Soldering a printed circuit board (PCB) involves working with high temperatures, electrical currents, and potentially hazardous chemicals, making it a potentially risky task. The following safety precautions should be taken while soldering PCBs to minimize the associated dangers.

Firstly, ensure a well-ventilated workspace or work in an area to prevent inhaling toxic fumes generated during soldering. Secondly, wear safety goggles and gloves to protect your eyes and hands from burns, molten solder, and accidental splashes of chemicals. Ensure that the soldering iron and associated equipment are properly grounded to avoid electric shocks. Additionally, it is essential to work on an insulated surface to prevent electrical shorts, and to keep flammable materials away from the work area to avoid fire hazards. Finally, always be mindful of the hot soldering iron and its cord to prevent burns or tripping accidents. By adhering to these safety precautions, the risks associated with soldering PCBs can be significantly minimized.

Soldering Safety Precautions

1. Always wear protective gear such as safety glasses, gloves, and long sleeves when soldering. Long hair should be tied back.

2. Adult supervision is required at all times.

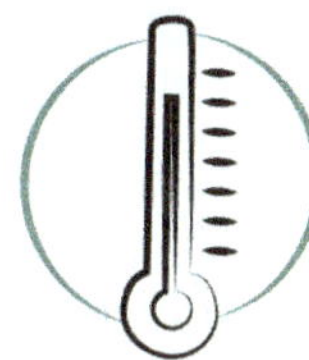

3. Make sure that the soldering iron is the correct temperature for the job.

4. Make sure that the soldering iron is not in contact with any flammable materials, or other objects, such as your skin or clothes.

5. Stand the soldering iron upright in a heat-resistant holder when not in use to avoid possible accidents. Make sure the cord is out of the way, and not in a place that could become a trip hazard.

6. Do not touch the tip of an iron with your bare hands.

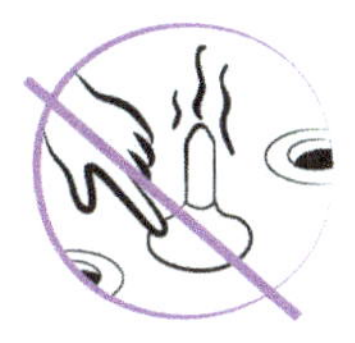

7. Do not touch any soldered parts before they have cooled.

8. Always work on a thermal, heat-resistant surface, like silicone, stone, or ceramic.

9.   Dispose of solder and flux properly.

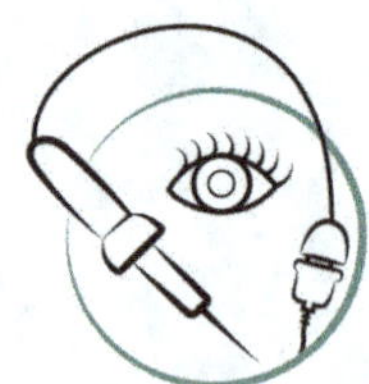

10.   Do not leave the soldering iron plugged in while unattended.

11.   Use a grounded outlet and always turn off and unplug the soldering iron when not in use.

12.   Have a fire extinguisher or bucket of sand, and first aid kit nearby when the soldering iron is in use.

13.   Be aware that Lead (Pb) is used in some solder. Avoid any contact with lead and be sure to clean up any lead dust or residue after soldering. It is safer, and therefore recommended, to use Lead-Free solder which is 99% Tin (Sn).

14.   Do not breathe smoke from soldering and work in a well-ventilated area.

15.   Do not touch your face while working with solder, and wash your hands after touching solder.

Check On Learning: Safety

1.   *What protective gear should you use, or what safety steps you should take when soldering? Check all that apply.*

   □ *Gloves*

   □ *Safety glasses*

   □ *A welding helmet*

   □ *Long sleeves*

   □ *Tie back long hair*

   □ *Grounding device*

   □ *Cold water*

   □ *Heat resistant holder*

   □ *Ear plugs*

   □ *First aid kit*

   □ *Fire extinguisher*

2.   *What kind of contact with lead can be harmful? Check all that apply.*

   □ *Inhalation, breathing in the smoke*

   □ *Skin contact*

   □ *Ingestion, getting it in your mouth or eyes*

3.   *What type of surface is heat-resistant? Check all that apply.*

   □ *Wood*

   □ *Stone*

   □ *Plastic*

   □ *Ceramic Tile*

   □ *Stainless Steel*

   □ *Silicon*

Materials

1. Safety Glasses

2. Gloves

3. Long sleeves

4. Heat resistant surface

5. Lead-Free, Rosin-Core Solder Wire, gauge .5mm to 1.5mm

6. Electronics Soldering Iron with Conical Tip (additional tips are a plus)

7. Heat-Resistant Stand for soldering iron

8. Sponge for cleaning tip or Brass Tip Cleaner

9. Practice Soldering Board: Printed Circuit Board (PCB)

10. Desoldering Pump

11. Desoldering Braid

12. Acid-Free Rosin Flux (Paste or Liquid)

13. Isopropyl Alcohol

14. Cotton Swabs

15. Small Paint Brush

16. Wire Cutters

17. Wire Stripping Tool

18. Needle Nose Pliers

19. Tweezers

20. Heat Shrink Tubing

21. Scissors

22. Insulated Wire, diameter 0.02 inches (0.5 mm) to 0.06 inches (1.5 mm)

23. Single Row Straight Header Strip Socket, 2.54 Mm Pin Heads (not shown)

24. Helping Hands, Soldering Tweezers or Soldering Clips: they act like a third and fourth hand to hold wires and components while soldering.

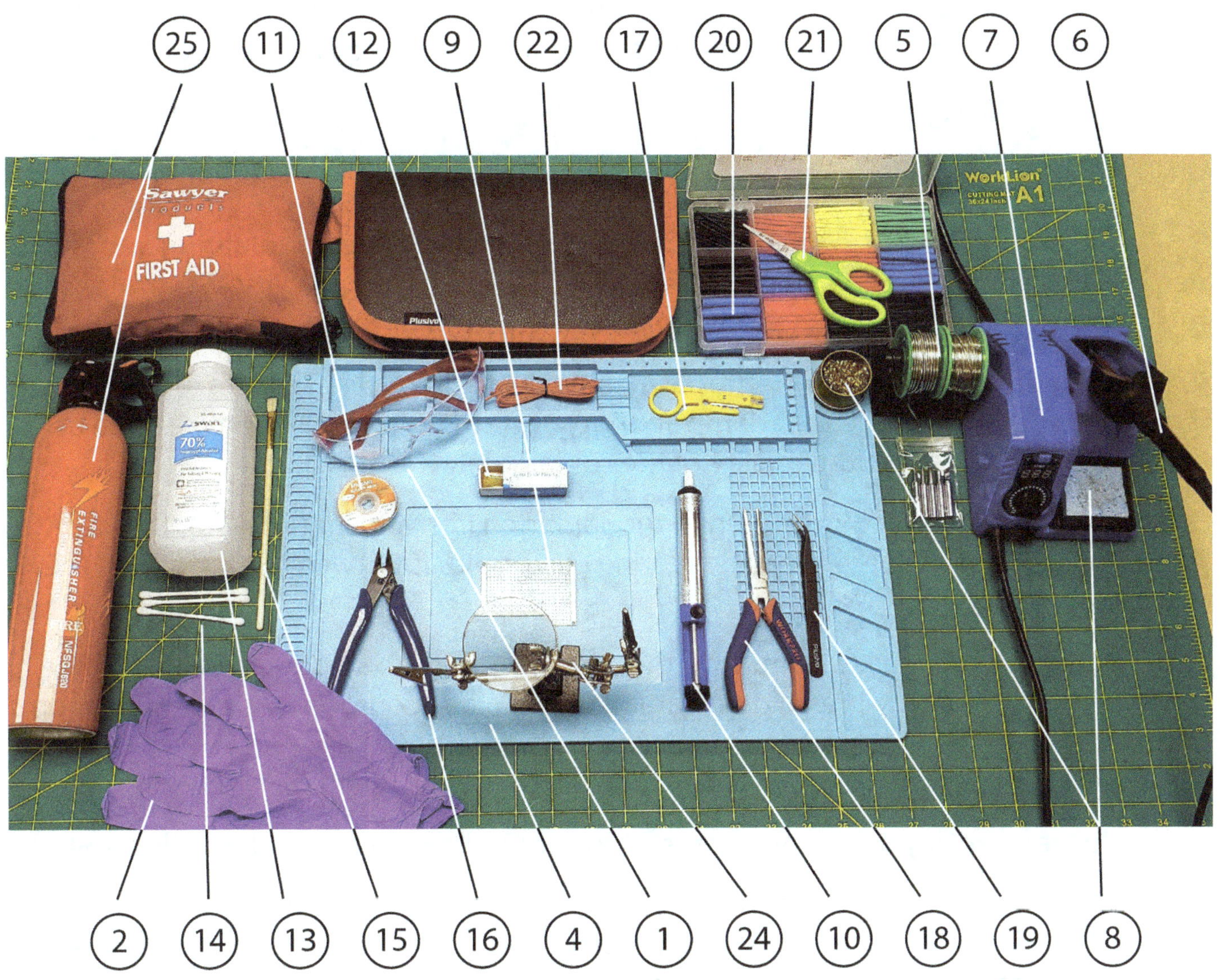

Soldering Hints

- Keep your sponge damp.

- Keep your tip clean.

- Keep fresh solder on your tip by Tinning it.

- Tin all wires and pads before soldering them together. Solder binds to solder more easily than it does to wire or other metals.

- Use the correct heat for the job.

- Use Rosin Core Solder for electronics (never use acid core).

- Use the shortest amount of time possible. Use high heat for a short amount of time, rather than low heat for a long amount of time. Circuit boards, components and pads are more likely to be damaged by longer exposure to low heat. High heat allows you to get in and get out quickly, thus limiting the length of exposure.

Soldering Step-By-Step

Here are the basic steps to follow for most soldering jobs.

1. **Gather the necessary tools and materials:** You will need a soldering iron, solder, heat resistant soldering stand. Depending on the job, you may also need wire cutters/strippers, flux, and the electronic component you want to solder.

2. **Clean the tip of the soldering iron:** Before you start, clean the tip of the soldering iron with a brass tip cleaner, damp sponge, or cloth. This removes any residue or oxidation that may affect the quality of the solder joint.

3. **Heat the soldering iron:** Plug in the soldering iron and let it heat up for a few minutes. The ideal temperature for soldering electronics is between 315°C and 370°C (600°F and 700°F).

4. **Prepare the area:** Cut and strip the wires you want to solder together. Clean glue or oxidation off any pre-soldered areas. Position any components in the appropriate area.

5. **Apply flux:** Apply a small amount of flux to the area where you want to solder. Flux helps remove any oxides on the metal surfaces, allowing the solder to flow better.

6. **Tin the area:** Apply solder to the end of each wire to "tin" them. Tinning coats the wire with a thin layer of solder, making it easier to solder the wires to the components. Do not apply too much solder or you can damage the wires.

7. **Solder the area:**

For Wires: Twist the wires together and hold them using pliers or a third hand tool, then touch the tip of the soldering iron to the joint. Apply a small amount of solder to the joint, making sure it flows evenly and covers the entire joint. Remove the soldering iron and let the joint cool for a few seconds.

For Components: Push the leads through the holes in the appropriate position on the PCB, and bend the leads on the under-side so it is secure. Turn the PCB over and touch a soldering iron to each lead for 1 to 3 seconds. Touch the solder to the opposite side of the lead, or wire, and allow the solder to melt and flow around the lead. Remove the solder and the iron. The solder should be in a mound shaped like a chocolate kiss, completely surrounding the lead or wire.

8.   **Finish the joint:** After the joint has cooled, use wire cutters to trim any excess solder, wires or leads. Inspect the joint to make sure there are no loose wires or cold solder joints, which are joints that don't conduct electricity, that could cause problems.

9.   **Repeat the process:** Repeat the process for any other wires or components you need to solder. Once you are finished, turn off the soldering iron and clean the tip with a damp sponge or cloth.

Note: Always follow safety precautions when soldering, such as wearing eye protection, keeping the soldering iron away from flammable materials, and unplugging the iron when not in use.

Good & Bad Soldering:

Good: Use the iron to heat the area, and then lay the solder on the area and let it melt in, to fill the joint.

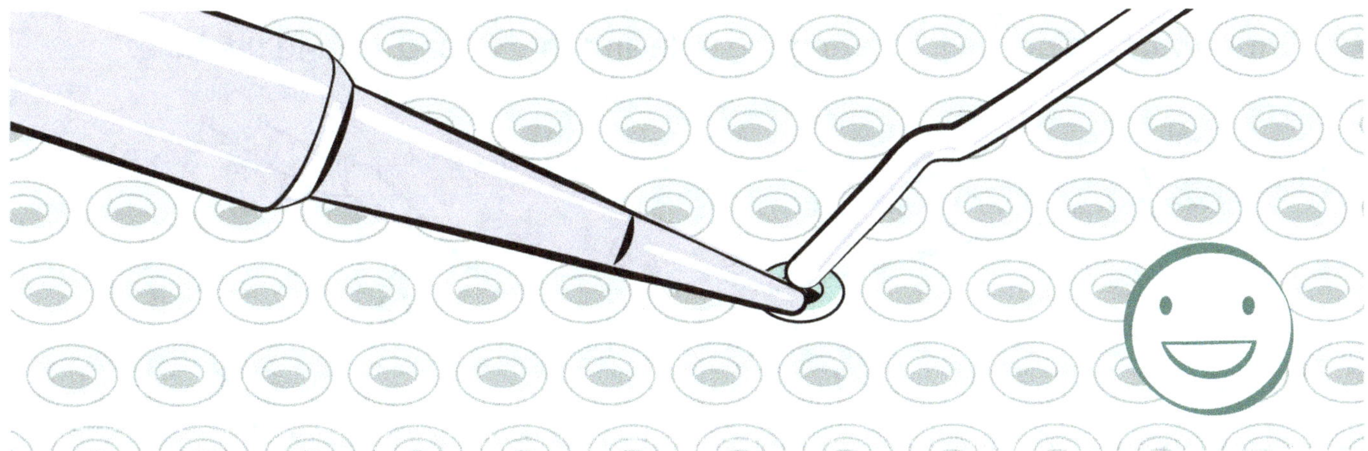

Bad: Don't melt the solder on the iron and try to pile or drip it on top of the area. This will lead to a messy board and possibly a "Cold" joint that doesn't conduct electricity. The point of soldering is to fuse the metals together and make a solid joint, and a good stable connection.

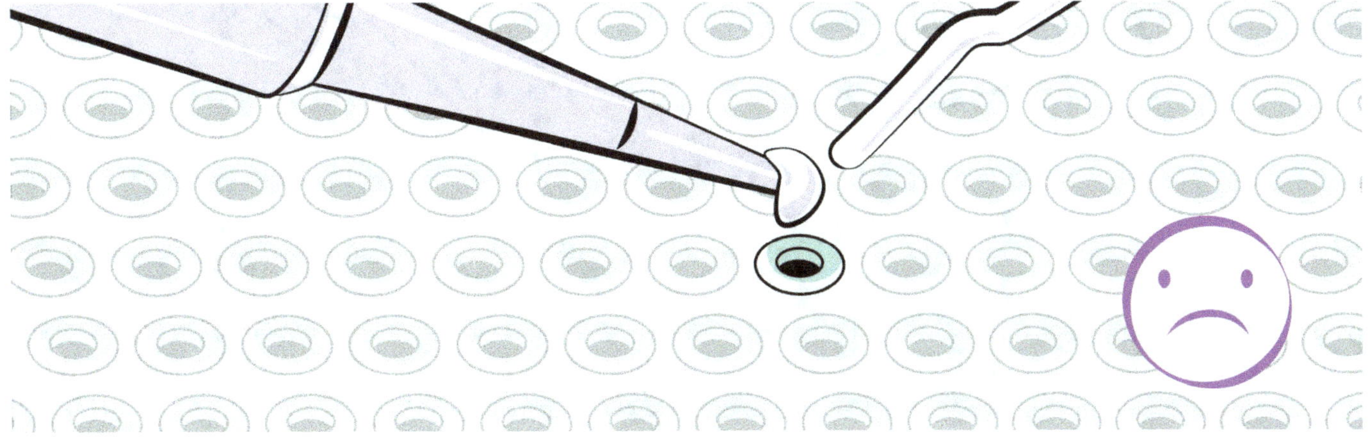

Good: Less is More—use just enough solder to make the connection. If you use too much, you run the risk of bridging a connection to neighboring components, creating a short circuit.

Good: High round solder ball, covering the hole and the "O" pad around the hole.

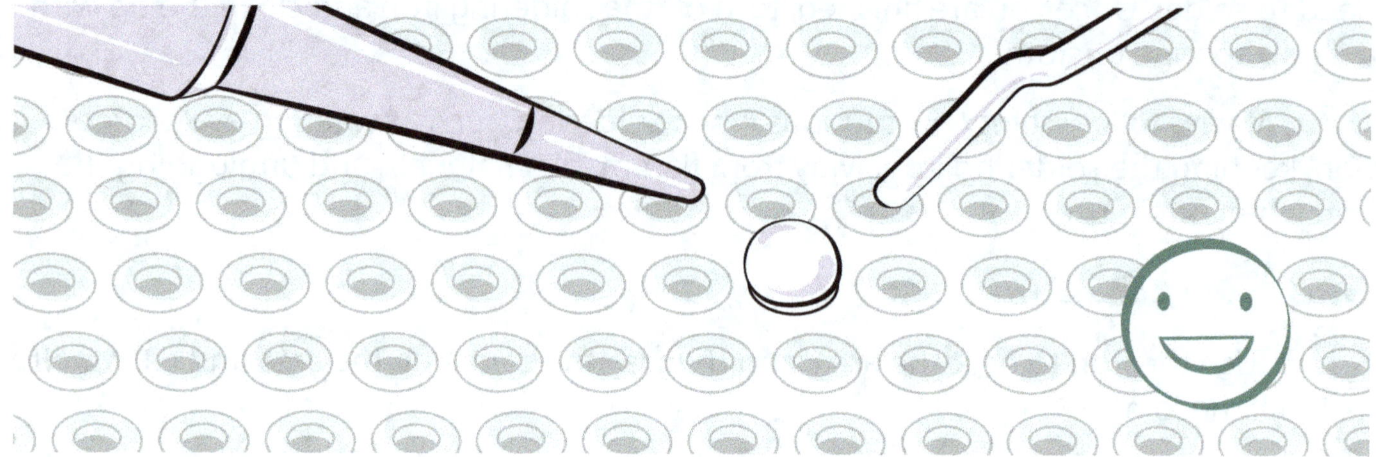

OK: Short round solder ball, covering the hole and the "O" pad around the hole.

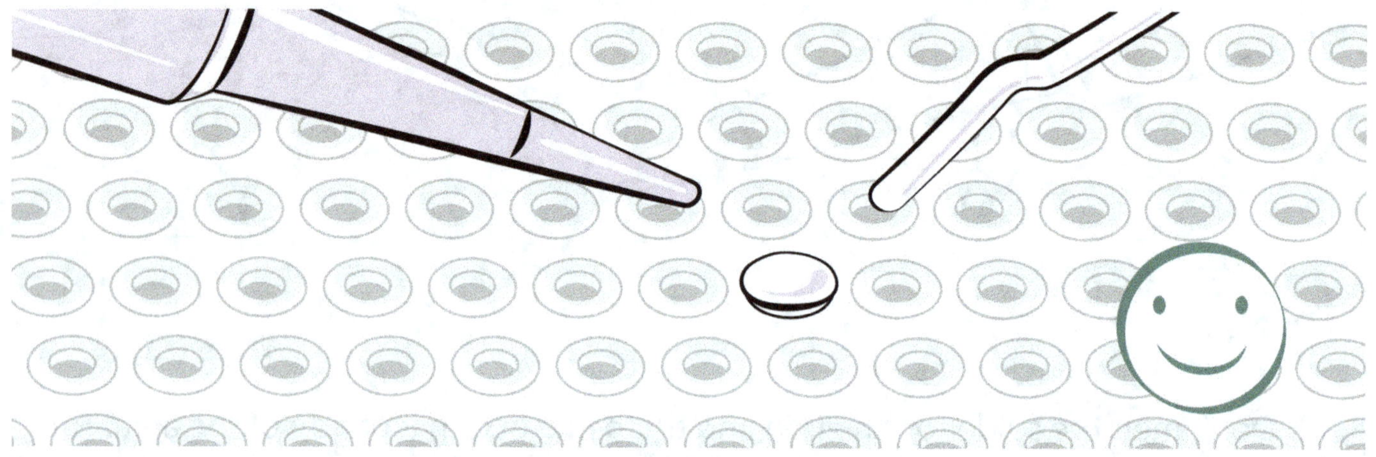

Bad: Not enough solder = Flat solder, with an indentation where the hole is.

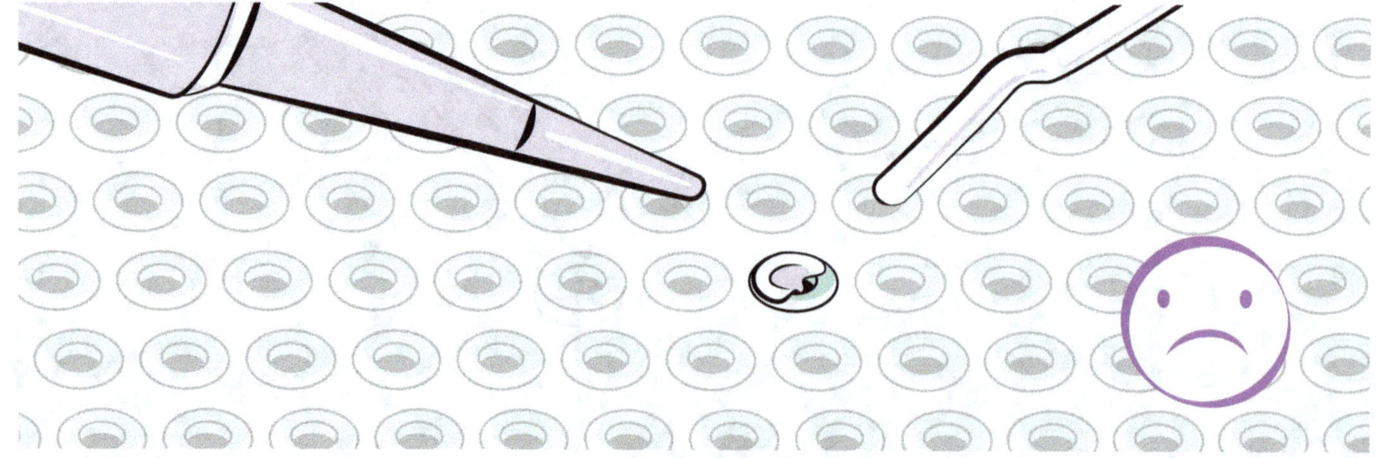

Bad: Too much solder = The solder flows outside the "O" and touches the hole next to it.

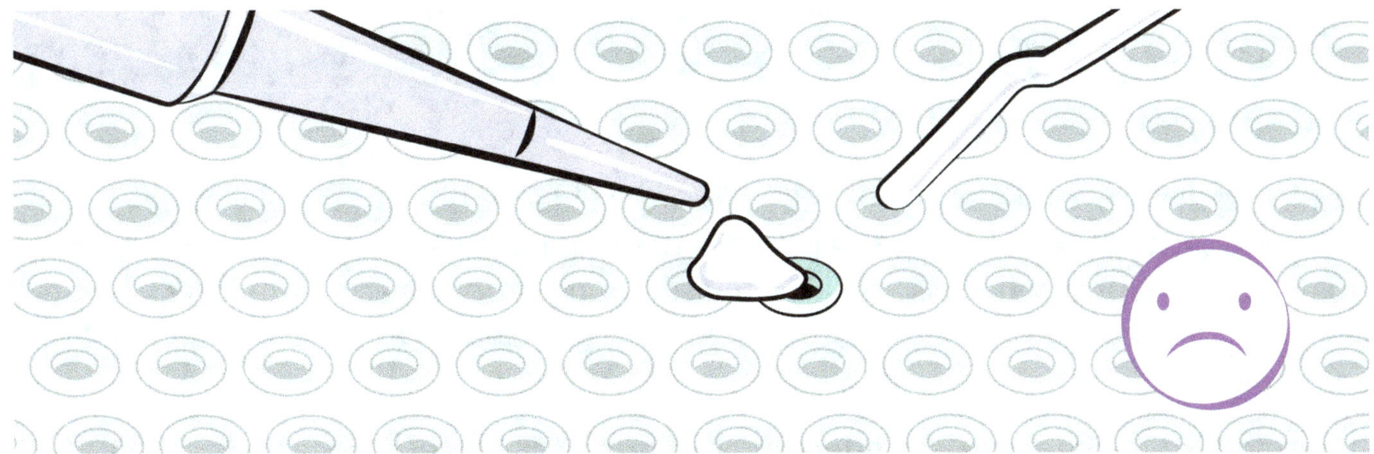

Check on Learning: Soldering

1.   *Why should you clean the tip of the soldering iron before starting?*
   a.   *To remove residue or oxidation that may affect the quality of the solder joint*
   b.   *To remove excess solder since old solder can't be re-used and won't flow to the solder joint*
   c.   *To prevent corrosion of iron tip due to rosin flux left on it for some time*

2.   *It is best to solder at high temperatures for a short period of time.*
   a.   *True*
   b.   *False*

3.   *Which of the following is within the ideal temperature for soldering electronics.*
   a.   *100 C*
   b.   *237 C*
   c.   *327 C*
   d.   *315 F*

4.   *You place the solder on the circuit board before adding heat.*
   a.   *True*
   b.   *False*

Lab 1: Tinning

Objective

Tinning is the process of applying a thin layer of solder to the surface of a metal, typically copper, before soldering. The purpose of tinning is to improve the solder joint's quality and strength while also making the soldering process easier. Tinning assists in removing oxidation from the metal surface, making it cleaner and more receptive to soldering.

It's important to keep your soldering iron tips tinned to protect them from oxidation during storage, and spread heat evenly while soldering. Tinning allows better heat transfer between the iron and the solder and results in a more reliable and consistent joint when soldering.

Tinning forms a protective layer of tin around the iron which helps to keep the heat in and protect the metal from oxidation. Oxidation is the process by which the metal acquires a dark color and becomes brittle. By tinning the tip of the iron, you help to prevent oxidation from happening and prolong the life of the soldering iron.

In electronic work, tinning is frequently used to tin the ends of wires or component leads before attaching them to other components.

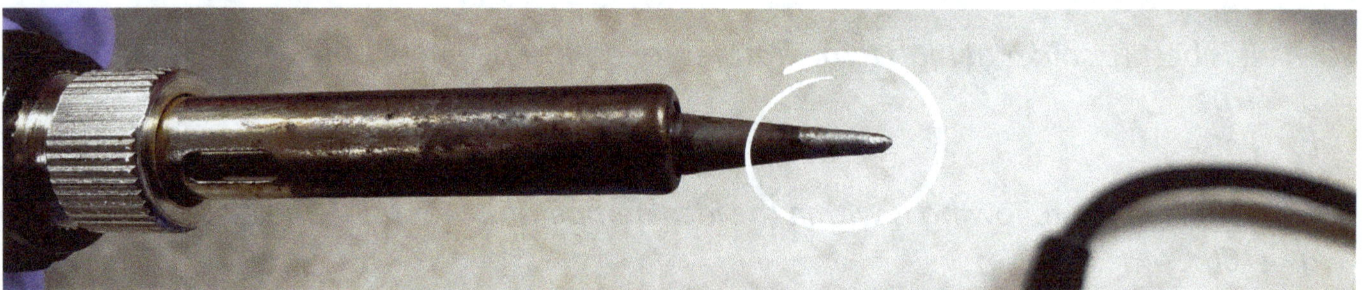

Materials

In addition to your personal protective safety gear, gather the necessary tools and materials you will need to do this lab.

- Soldering Iron with a Conical Tip

- Soldering Iron Stand

- Lead-Free Rosin-Core Solder

- Damp Sponge

Instructions

It is vital to use the correct amount of solder. When tinning apply a smooth thin layer of solder and avoid creating globs of solder that could cause electrical shorts.

1.   Dampen the sponge with water.

2.   Before you plug it in, make sure the handle and tip sleeve of soldering iron are on tightly.

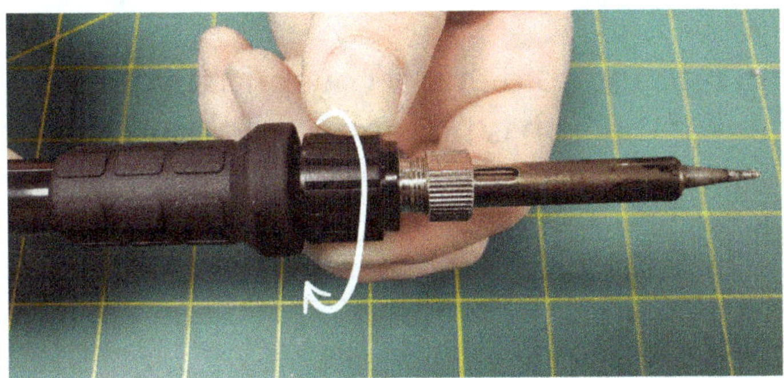 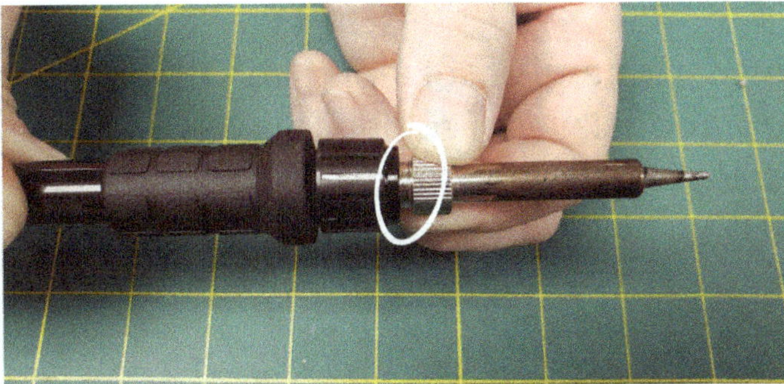

3.   Plug in the soldering iron and set it to the correct temperature (650 - 700 degrees Fahrenheit/343 - 371 degrees Celsius).

4.   Hold the soldering iron like a pencil, with the tip pointing away from you and your body. Do not touch or grasp the iron below the heat resistant handle when it is plugged in.

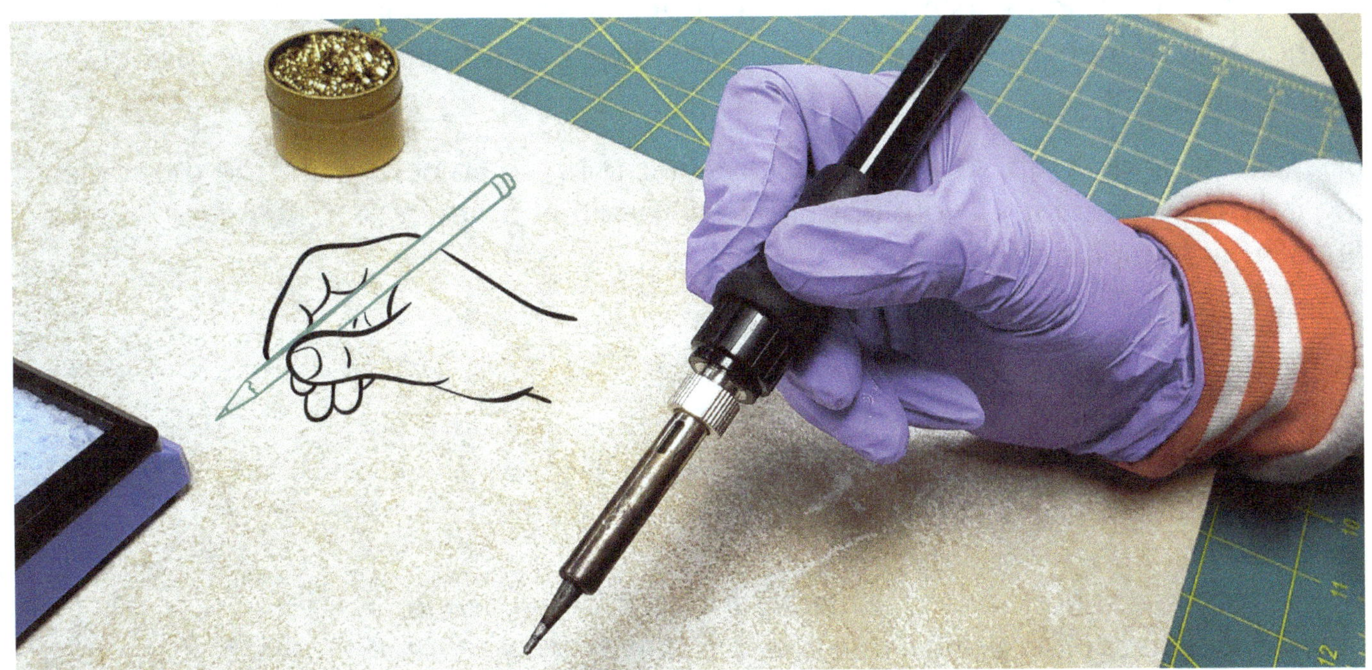

5.   Clean the soldering tip by wiping it on the damp sponge.

6.   Tin end of soldering iron tip by touching it to the solder and covering the tip with a smooth layer of solder. This will also determine if the iron is hot enough to melt the solder.

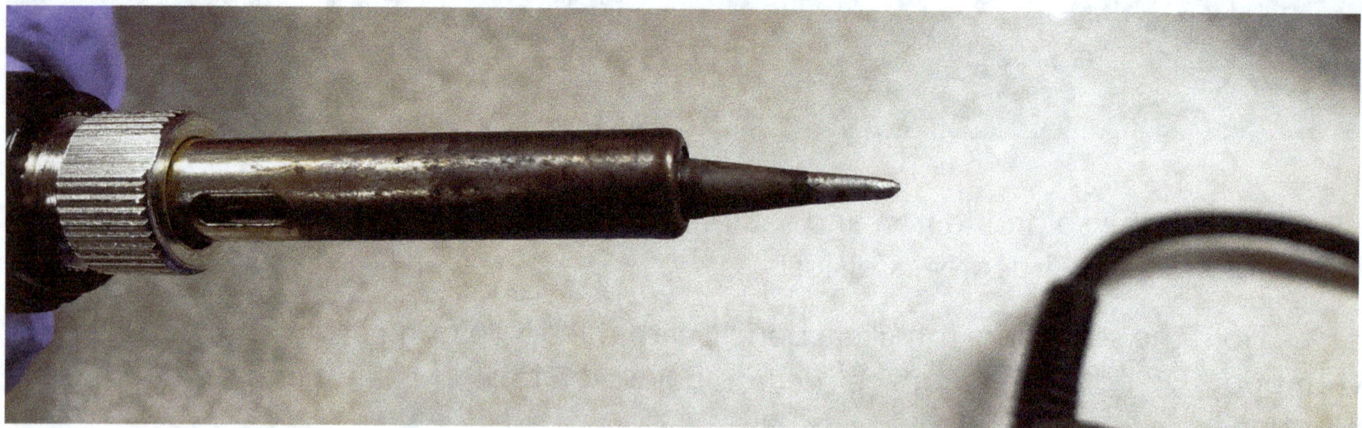

7.   Take a few moments now, to check your soldering iron's instructions and learn to change tips. Remember to unplug your soldering iron and allow each one to cool before touching the tip.

8.   After doing a job, you should always tin the tip again before putting the soldering iron away. Keeping all your tips tinned will protect them from rust and extend the life of your equipment.

Check on Learning: Tinning

1.   *What is tinning?*
   a.   *The process of cleaning the metal surface before soldering.*
   b.   *The process of applying a thin layer of solder to the surface of a metal before soldering.*

c.   *The process of heating the soldering iron to a high temperature before soldering.*
d.   *The process of using too much solder while soldering.*

2.   *What is the purpose of tinning?*
a.   *To make the metal surface brittle.*
b.   *To make the soldering process quicker.*
c.   *To prevent oxidation of metal surfaces.*
d.   *To create globs of solder on the metal surface.*

3.   *How should you hold the soldering iron?*
a.   *Like a hammer, with the tip pointing towards you.*
b.   *Like a pencil, with the tip pointing away from you and your body.*
c.   *Like a toothbrush, with the tip pointing towards your face.*
d.   *Like a spoon, with the tip pointing downwards.*

4.   *It is suitable to clean the tip of a hot soldering iron on:*
a.   *Paper Towel*
b.   *Damp Sponge*
c.   *Dry Sponge*
d.   *Desoldering Braid*
e.   *Your Sleeve*

Lab 2: Soldering a Practice PCB

Objective

Soldering is a skill that takes practice to master. It's important to practice soldering electronics in order to develop the necessary skills and techniques required to make reliable connections between electronic components. Through practice, a person can learn how to properly heat the components, melt the solder, and create a strong, lasting bond. Achieving this level of proficiency is necessary for ensuring that any electronics created are safe, reliable, and functional.

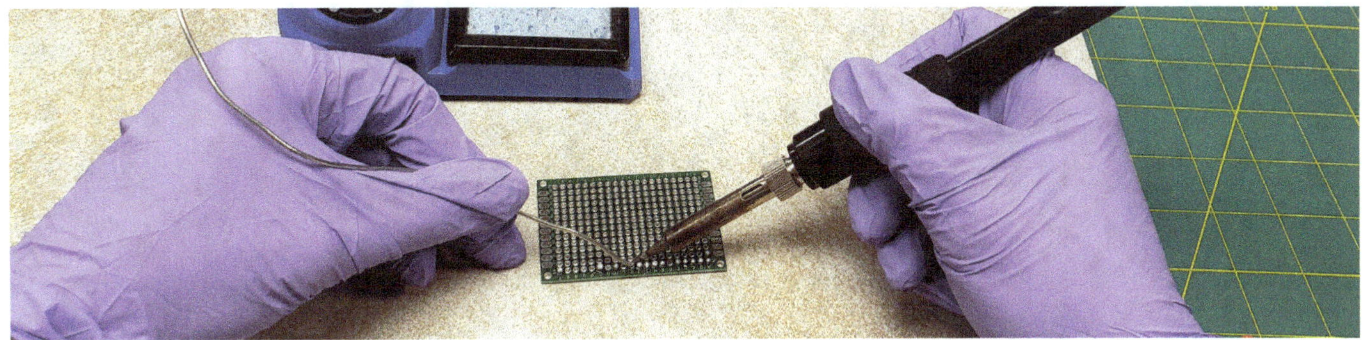

Materials

In addition to your personal protective safety gear, gather the necessary tools and materials you will need to do this lab.

- Soldering Iron with a Conical Tip and a Knife Tip

- Soldering Iron Stand

- Lead-Free Rosin-Core Solder

- Damp Sponge

- Practice PCB board

Instructions

1. Gather your supplies and set up in a well ventilated area.

2. Dampen the sponge with water.

3. Start by putting the Conical Tip on your cool soldering iron. Before you plug it in, make sure the handle and tip sleeve of soldering iron are on tightly.

4. Plug in the soldering iron and set it to the correct temperature (650 - 700 degrees Fahrenheit/343 - 371 degrees Celsius).

5. Taking the iron in your dominant hand, apply heat to the Practice PCB by placing the tip of the iron on one side of the metallic silver or copper "O" pad surrounding a hole on the circuit board.

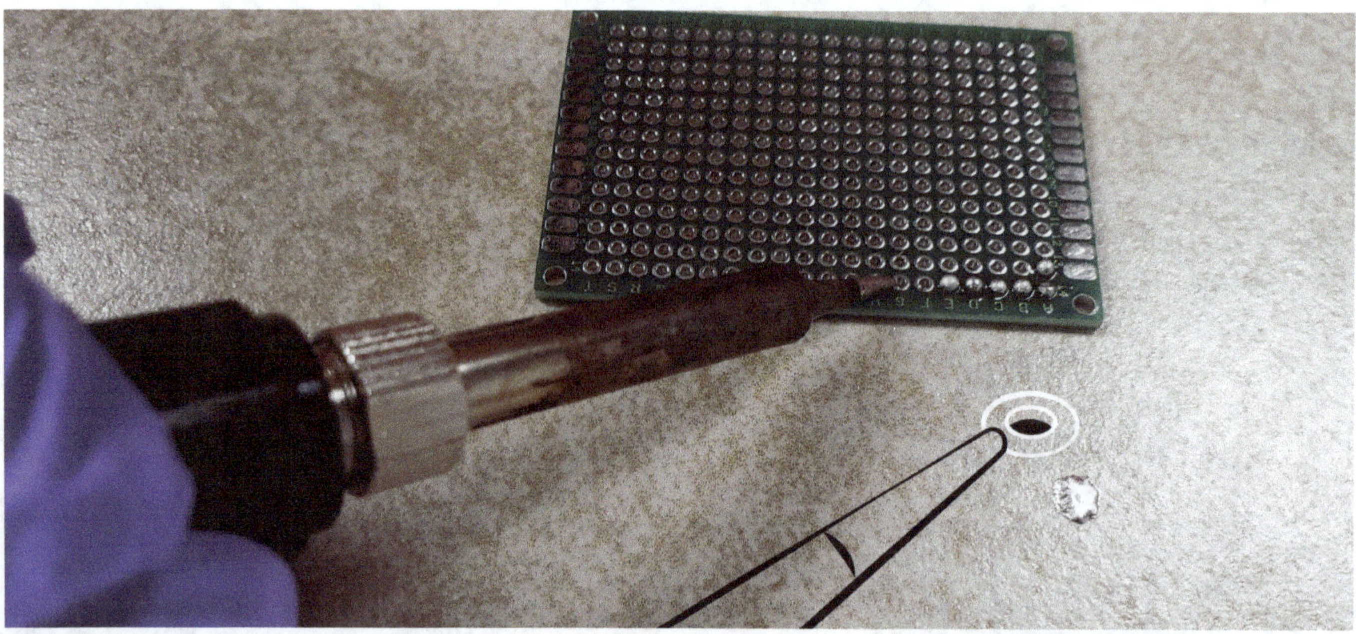

6.   Hold it there for about one to three seconds to heat it up.

7.   Pick up the solder wire in your other hand, holding it 3 to 4 inches away from the end. Make sure to keep your fingers far away from the end of the solder wire, because when it starts to melt it will disappear quickly, bringing your fingers closer to the soldering iron.

8.   Bring the solder to the "O" pad and place the solder wire onto the heated area and allow it to melt and flow into the area, creating a mound or ball that completely covers the hole.

9.   Remove the soldering iron and the tip of the soldering wire at the same time.

10.   Wait two to three seconds for the solder to cool. When cool, it will change from shiny silver, to a dull silver, as seen on figure above.

11.   Inspect the Solder to ensure solder balls do not touch each other, or you will have a short circuit. If it doesn't look right, you may need to remove and re-solder the ball.

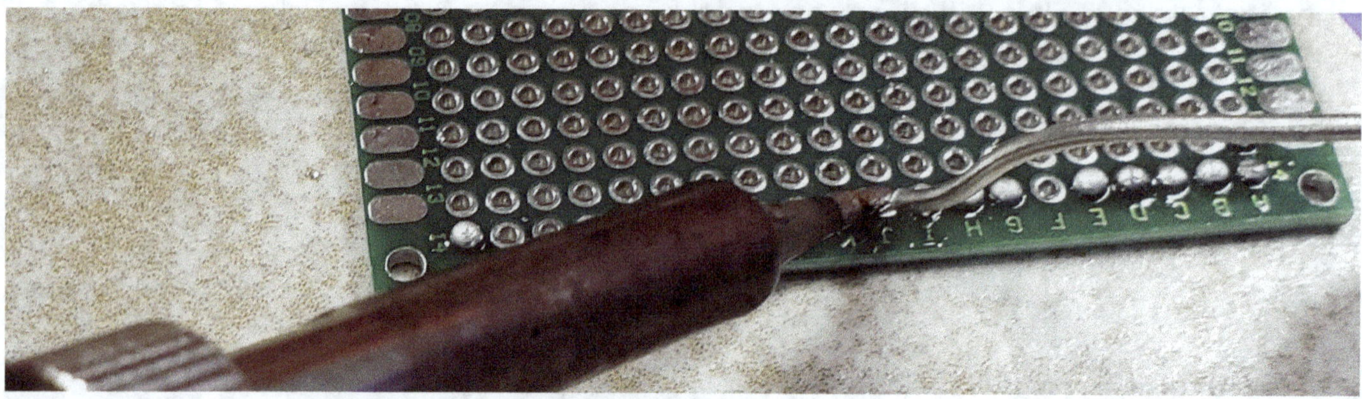

12.   Keep practicing until you are consistently creating high round balls of solder that do not touch each other.

Using the Knife Tip

When you get good at the Conical Tip, it's time to practice with the Knife Tip. This is a versatile tip that can offer different types of soldering by using either the edge or the tip of the blade.

1.   Unplug the iron and allow it to cool.

2.   Remove the Conical Tip and install the Knife Tip making sure the handle and tip sleeve of the soldering iron are on tightly before plugging it back in.

3.   Hold the iron with the pointy tip of the knife facing down, towards the PCB.

4.   Practice soldering using the same method as before until you feel comfortable using the tip of the knife tip as you would the conical tip.

5.   Notice how you can easily use the flat part of the knife blade by rotating the iron, effectively creating a new tool without switching tips. This is helpful when doing precise and detailed work such as cutting small wires, cleaning out solder from tight spaces, or scraping away excess solder to make the joint look cleaner. The knife tip is ideal for working in confined spaces or intricate soldering jobs.

Check on Learning: PCB Board

1.   *What should you do after putting the soldering iron against the metallic "O" pad on the circuit board?*
   a.   *Pick up the soldering iron and move on to the next pad*
   b.   *Hold the soldering iron in place for 10 seconds*
   c.   *Wait for 1 to 3 seconds before proceeding*
   d.   *Apply solder to the "O" pad immediately*

2.   *How can you tell if the solder has cooled and solidified?*
   a.   *It will turn from shiny silver to dull silver*
   b.   *It will start to smoke*
   c.   *It will melt and flow into other areas*
   d.   *It will start to bubble*

3.   *What should you do if the solder balls touch each other?*
   a.   *Continue practicing until you get it right*
   b.   *Leave it as is, it won't cause any issues*
   c.   *Add more solder to the ball*
   d.   *Increase the temperature of the soldering iron*

4.   *What should you do when practicing soldering?*
   a.   *Connect the solder balls to the circuit board with super glue*
   b.   *Hold the solder wire 3 to 4 inches from its end*
   c.   *Add heat using a heat gun*
   d.   *Put on a welding helmet to protect your eyes*

Lab 3: Soldering Wires Together

Objective

Let's say, for example, you have a drone that is powered by a battery, and the wires that connect the battery to the motor get disconnected or pulled apart. To fix it, you need to solder the two wires together. Soldering helps to create a strong, permanent connection between the two wires.

As a fun project, in this lab we will solder the 2 ends of one wire together to make a bracelet or necklace.

Materials:

In addition to your personal protective safety gear, gather the necessary tools and materials you will need to do this lab.

- Soldering Iron with a Conical Tip and a Knife Tip

- Soldering Iron Stand

- Lead-Free Rosin-Core Solder

- Damp Sponge

- Wire cutters/strippers

- Flux

- Heat shrink tubing (or electrical tape)

- Scissors

- Helping Hands, Soldering Tweezers or Soldering Clips

- For this project you will also need insulated wire. The diameter of wire used in electronics typically ranges from 0.02 inches (0.5 mm) to 0.06 inches (1.5 mm).

- Hair Dryer or Heat Gun

Instructions:

1.   Cut a piece of wire to the desired length (7" - 12" for a bracelet; 25" - 30" for a necklace).

2.   Use a wire stripper to remove the insulation off the ends of the wire, to uncover approximately 1/4 inch (.5 cm) of bare wire. Make sure the stripped ends of the wire are clean and free from any debris or insulation.

3.   Put on the Heat Shrink Tubing: use scissors to cut the Heat Shrink Tubing to one inch (2.5 cm) longer than the twisted wire joint (approximately 2 inches/5 cm). Pass the wire through at least one heat shrink tube. If you're feeling creative, add several pieces of different colors of heat shrink tube. (Skip this step if you are using electrical tape.).

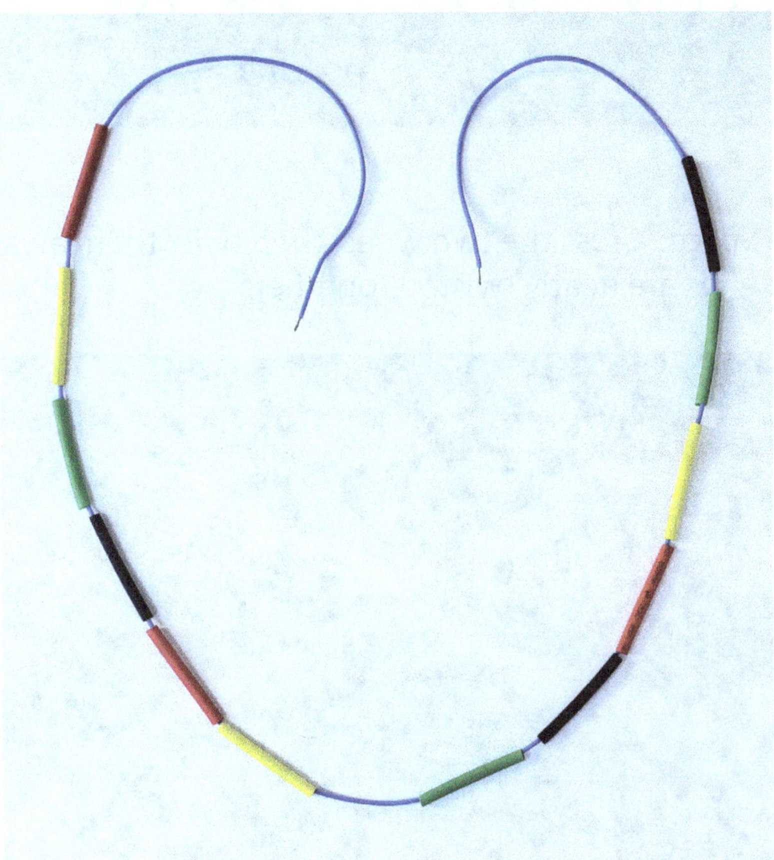

4.   If your wire has strands, there are 2 equally effective ways to twist, or splice, the two wires together.

   a.   The Fan Splice: Fan out the strands on both ends of the wire. Insert the fans into each other, interweaving the strands, and twist the bare ends of the strands together neatly.

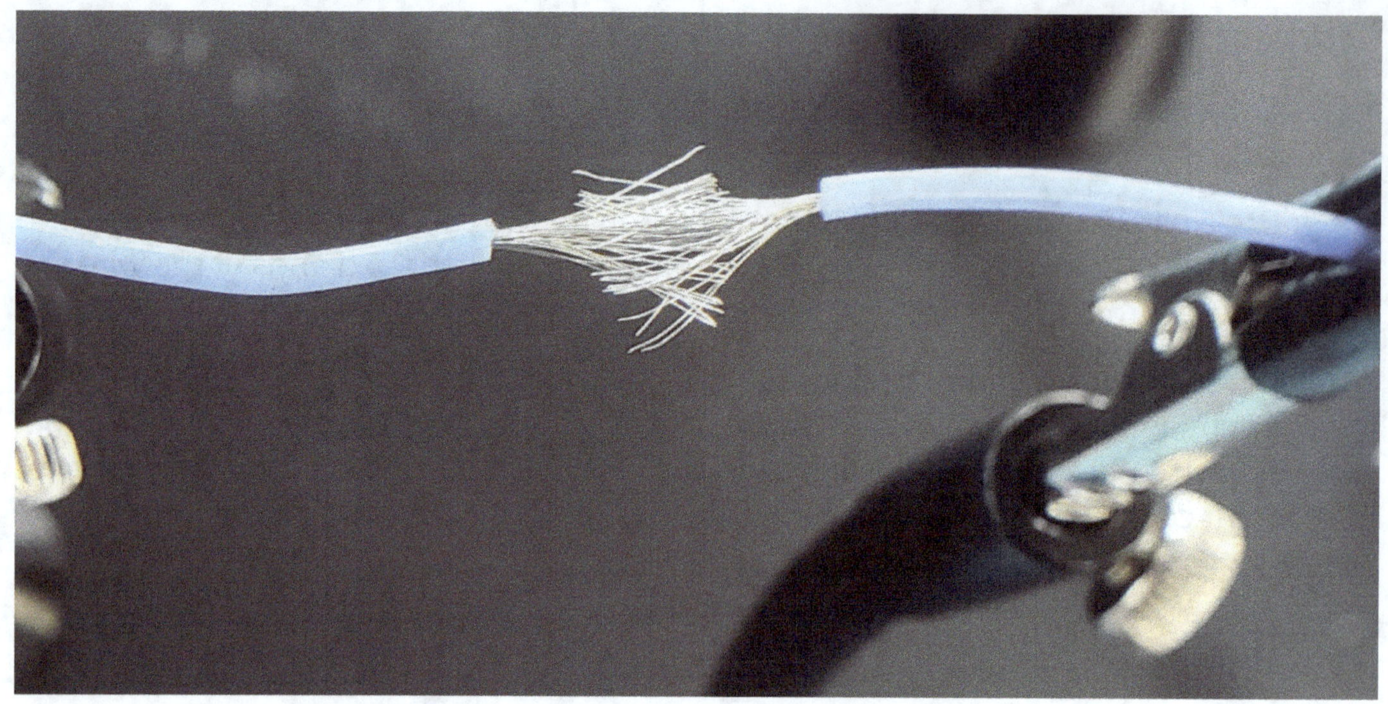

b. The Lineman Splice: Cross the two wires, then twist them away from each other. Make sure all the ends are neatly twisted together.

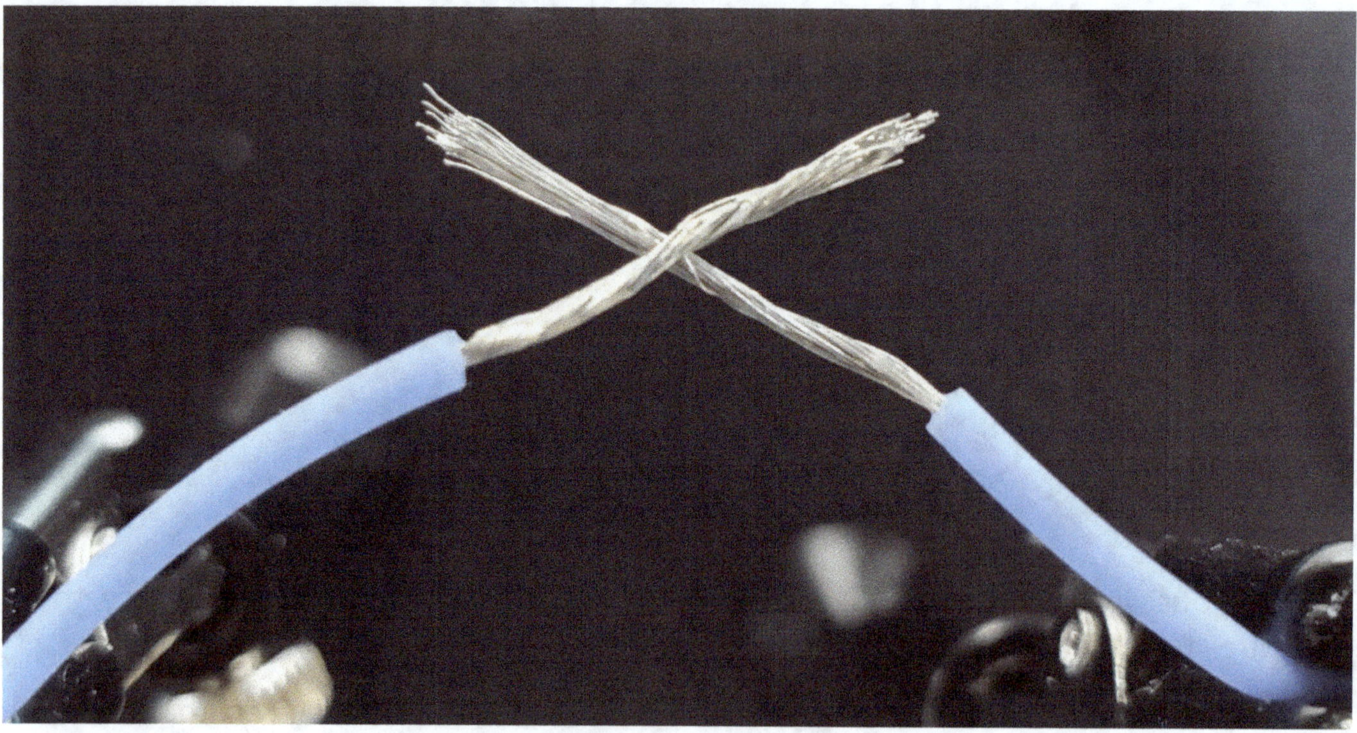

5.   Install the Conical Tip on the soldering iron and plug it in. Heat it to approximately 675 degrees fahrenheit (350 degrees celsius).

6.   Tin the tip of the soldering iron.

7.   Use Soldering Clips/Helping Hands to hold the wire as you work.

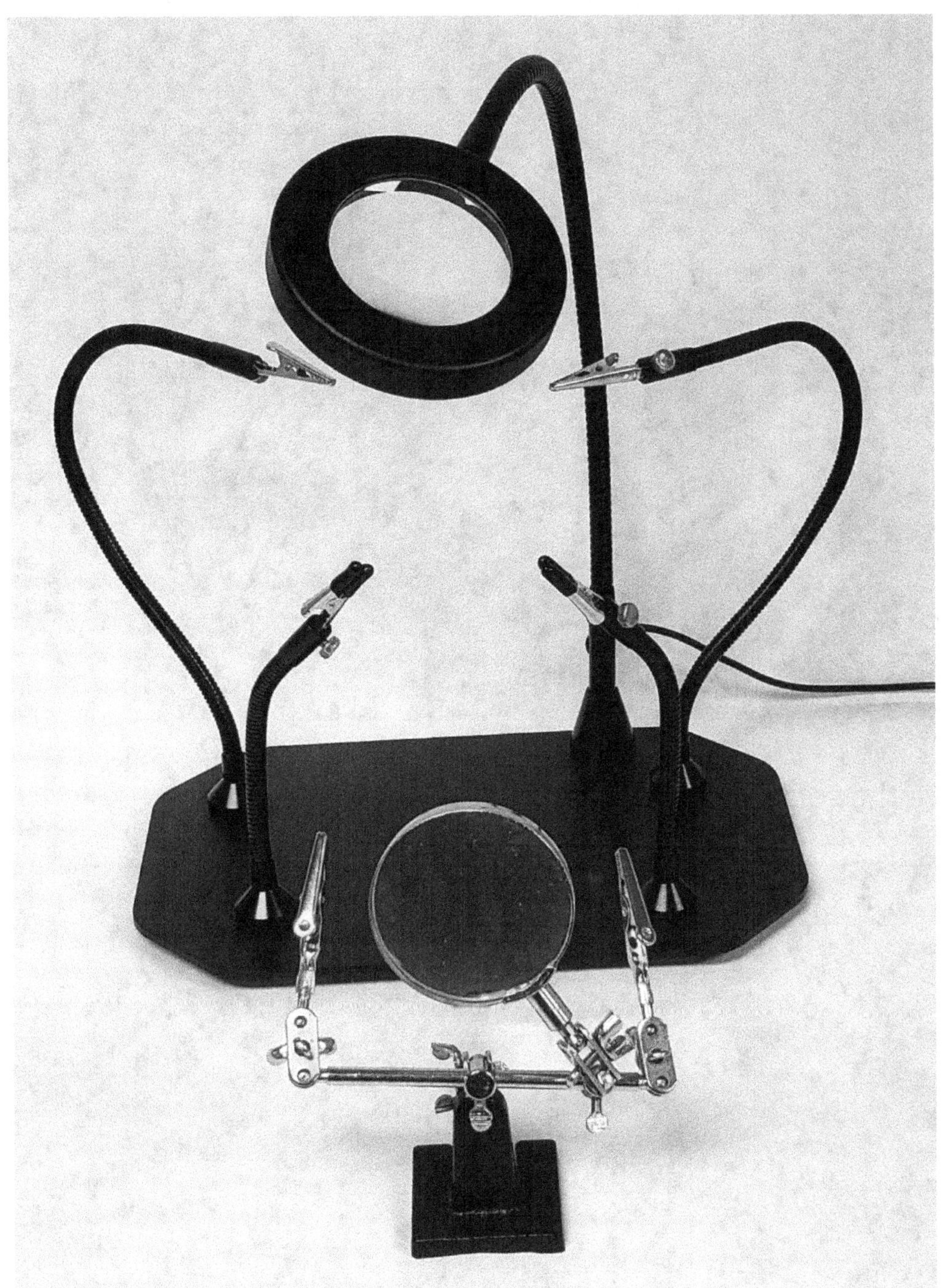

8.   Apply flux to the twisted joint of bare wire using a brush or swab. Flux helps remove any oxides on the metal surfaces, allowing the solder to flow better.

9.   Place the tip of the iron on the under-side of the twisted wire joint and allow it to heat up for 1 to 3 seconds.

10.   Touch the solder to the top-side of the twisted wires, allowing the solder to melt and flow over and into the entire joint. Note, there is no need to touch iron directly on the solder.

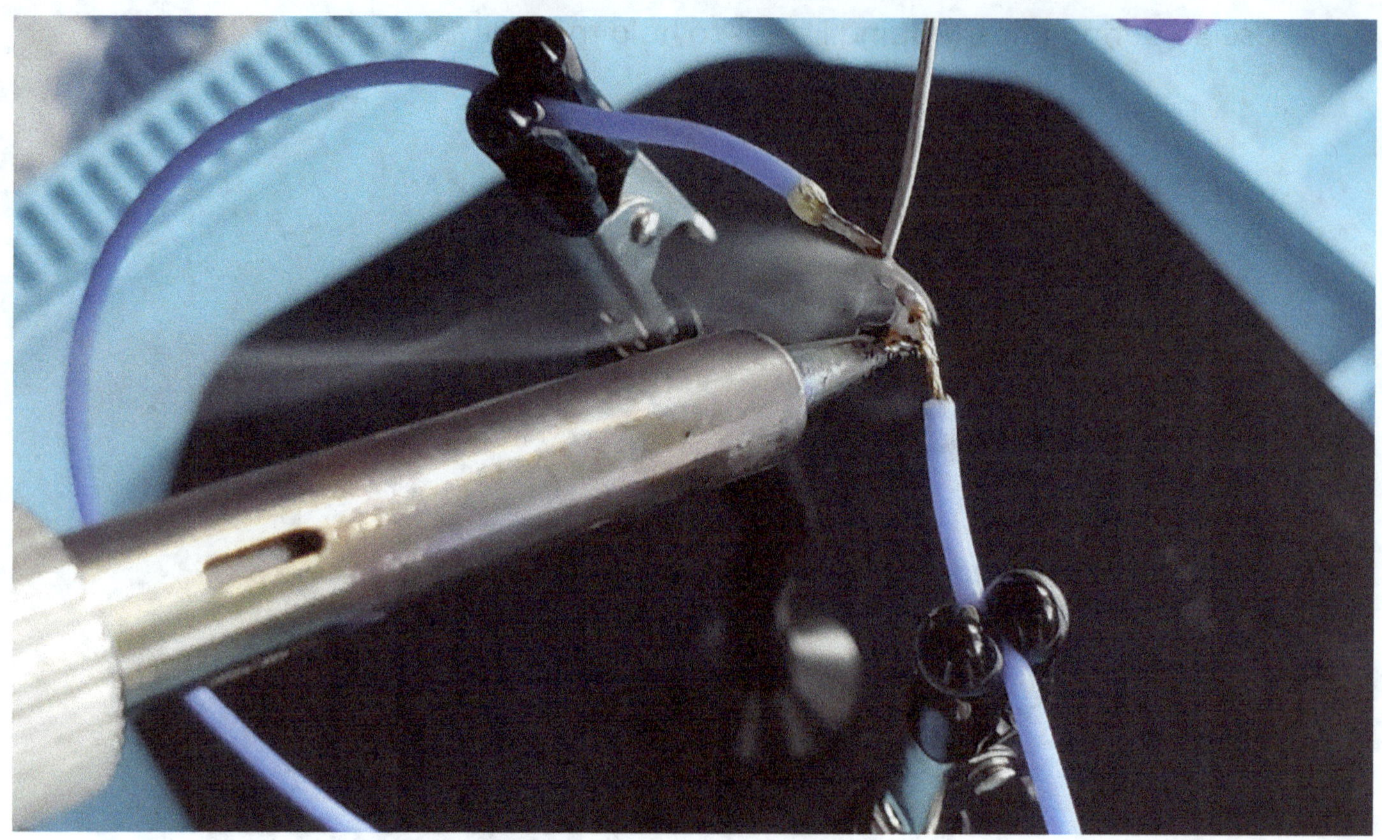

11.   Remove the soldering iron and allow the joint to cool.

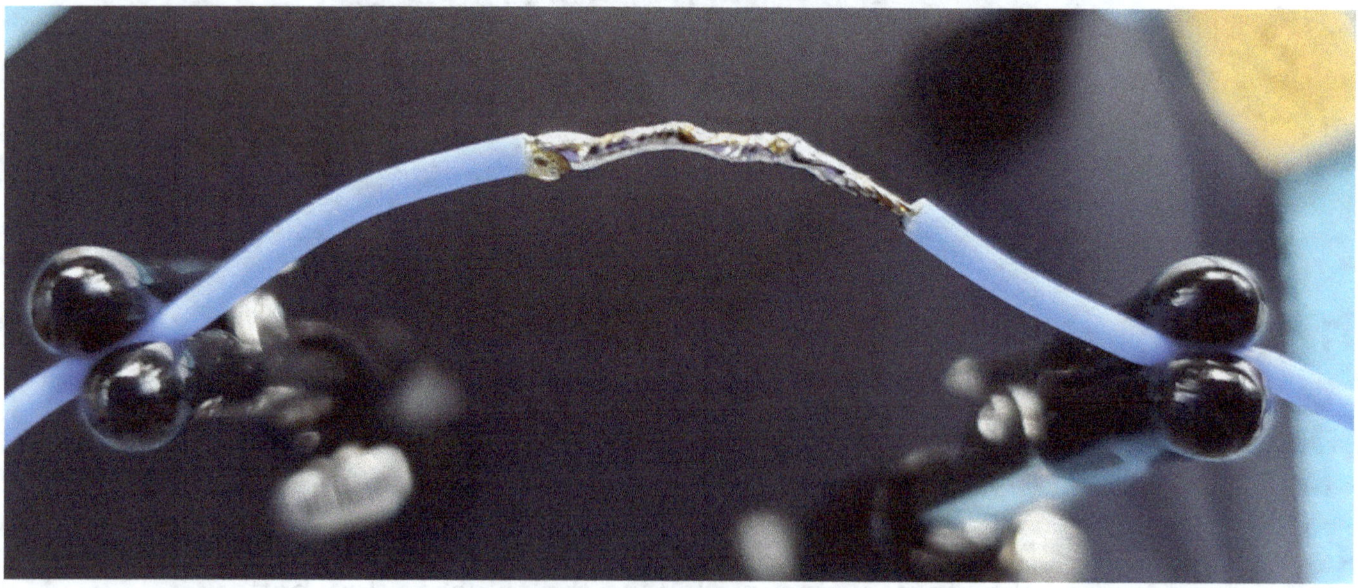

12.   Finish the joint: After the joint has cooled, inspect the joint to make sure it is secure and all the exposed wire is covered with solder. If necessary, reheat the joint and add or remove solder as needed.

13.   If a waterproof joint is desired, spread on Silicon Paste with a brush or swab. (Not shown.)

14.   Insulate the joint: Slide the heat shrink tube over the joint and heat it with a hair dryer (not all hair dryers get hot enough) or heat gun so it shrinks, covering all exposed wire. Heat the decorative tubes as well so they shrink to the insulated wire. If you are using electrical tape, wrap the tape tightly around the wire joint to cover all exposed metal. Similarly, you can use different colored electrical tape to decorate the necklace or bracelet as desired.

Congratulations, you just made a circle out of wire. You can use this technique with varying colored wires, lengths, and heat shrink tubes to make a necklace!

Check on Learning: Soldering Wires Together

1.   *What should you do before you splice the wires?*
   a.   *Tin the wires.*
   b.   *Put on the heat shrink tubing.*
   c.   *Twist the bare ends of the two wires together to create a circle.*
   d.   *Cut the wires to the desired length.*

2.   *What is the purpose of heat shrink tubing when soldering wires together?*
   a.   *To help the wires cool down*
   b.   *To make the wires easier to bend*
   c.   *To cover the exposed wires*
   d.   *To create a decorative effect*

3.   *What should you do after the joint has cooled?*
   a.   *Inspect the joints to make sure they are secure and all the exposed wire is covered with solder*
   b.   *Apply more flux to the joint*
   c.   *Cut off the excess wire with wire cutters*
   d.   *Leave the joint as is and move on to the next step*

Lab 4: Soldering Wires to a Circuit Board

Objective

Let's say you want to add a light to your drone, and you want to attach it to the circuit board. Practice soldering the tiny wires from the camera to a circuit board. In this exercise we'll practice soldering wires to a PCB.

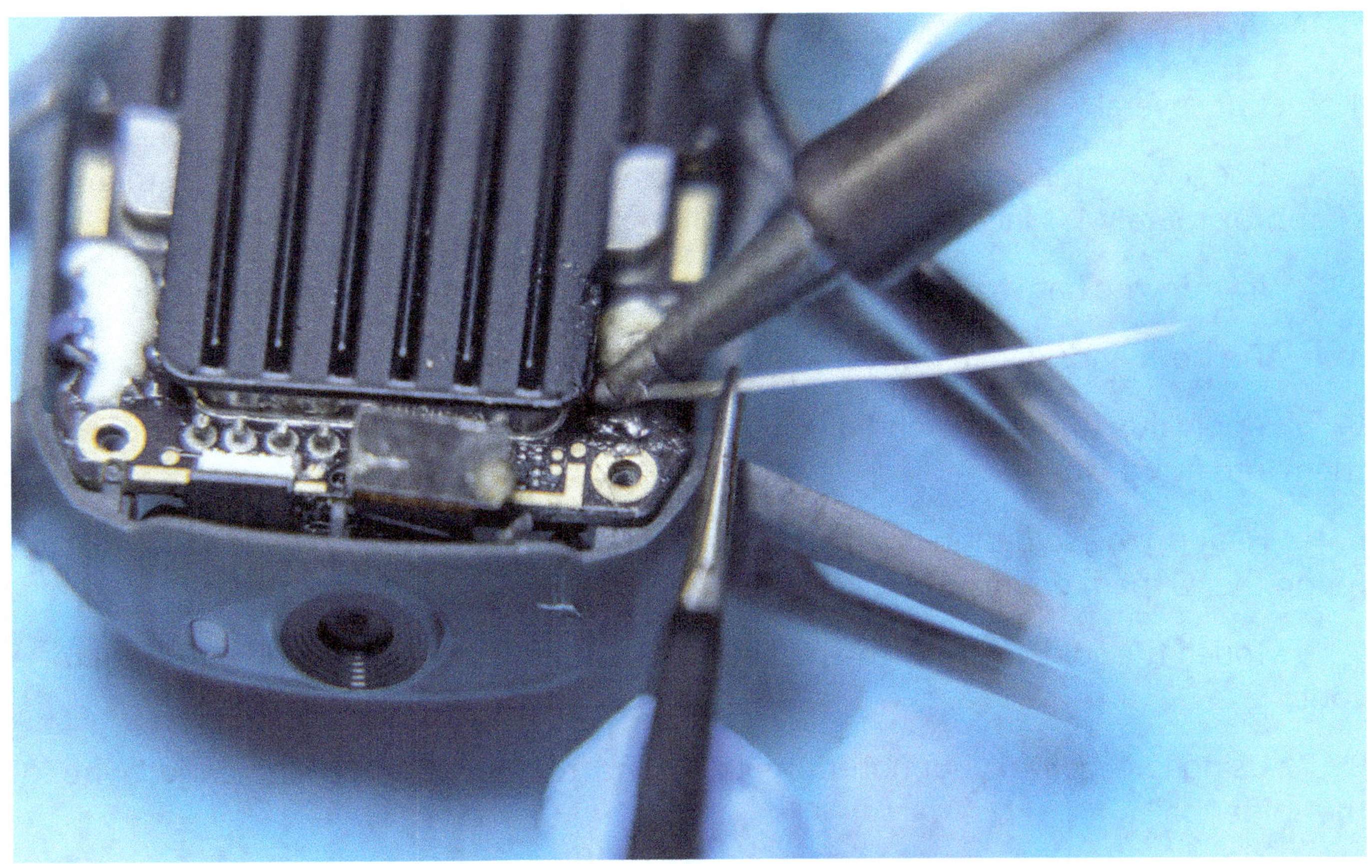

Materials

In addition to your personal protective safety gear, gather the necessary tools and materials you will need to do this lab.

- Soldering Iron with a Conical Tip and a Knife Tip

- Soldering Iron Stand

- Lead-Free Rosin-Core Solder

- Damp Sponge

- Wire cutters/strippers

- Several Pieces of Insulated Wire about 3" - 6" long (7.5 - 15 cm)

- Practice PCB with rectangular pads. (If you don't have one, use a Standard PCB with holes. For this exercise you will disregard the holes.)

- Acid-Free Rosin Liquid Flux and Small Paint Brush to apply Flux, or Flux paste

- Soldering Clips/Tweezers

Instructions

1. Cut several wires to the desired length.

2. Use a wire stripper to remove the insulation off both ends of each wire, to uncover approximately 1/4 inch (.5 cm) of bare wire.

3. If your wire has strands, twist the bare ends of the strands together neatly.

4. Tin both ends of each wire.

5. Apply flux to the bare end of wire using a brush (for liquid flux), or by dipping them into flux paste.

6. Heat up the soldering iron and place the tip of the iron on the rectangular pad (or on the "O" pad) on the PCB.

7. Touch the solder wire to the pad and leave a small solder ball. Allow the ball to cool.

8. Using tweezers in your non-dominant hand, touch one end of the tinned wire to the solder ball.

9. Heat the solderball with the iron and use the iron to press the wire into the melted solder ball. Try to press the wire all the way into the solderball, so that the tip is completely covered, and there is no bare wire exposed.

10. Remove the heat and hold the wire steady until it cools.

11. Inspect the joint to make sure it is secure. Make sure the solder does not flow beyond the pad, and that it does not come in contact with another solder joint. If you see any problems, try reheating the joint and fixing it.

12.   Continue soldering wires to the board, connecting both ends of wire around the board until you feel comfortable with this technique, and you are consistently creating strong secure wire-to-board connections.

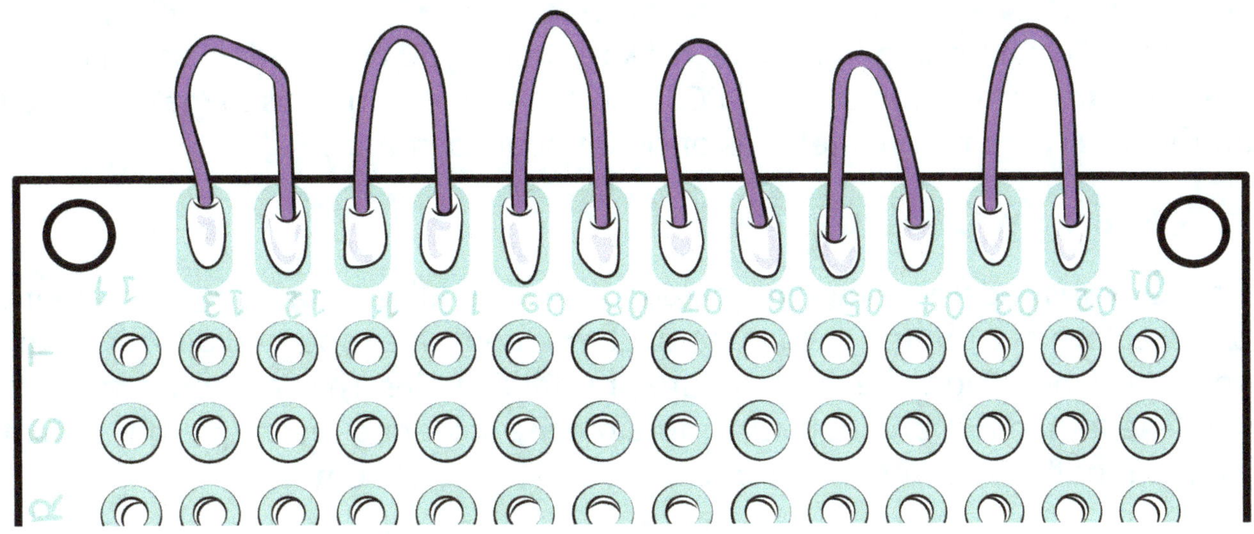

Congratulations! You've just completed your first Surface Mount. To find out what that means, read on….

Check on Learning: Surface Mount

1.   *What should you do if you see any problems with the solder joint?*
   a.   *Ignore it and move on to the next joint*
   b.   *Heat the joint and try to fix it*
   c.   *Leave it alone and hope it doesn't cause any issues*
   d.   *Cut the wire and start over*

2.   *True or False: You should see a bare wire sticking out of the solderball.*
   a.   *True*
   b.   *False*

3.   *What three things should you do to each end of wire before soldering?*
   a.   ____
   b.   ____
   c.   ____

In order to build something from scratch, like a drone, you will need to attach many different electronic components to a PCB. There are two main ways to make a circuit board that use two different methods of circuit board assembly.

Surface Mount Device (SMD)

This is the more modern method of soldering which involves mounting components onto the surface of a printed circuit board. Components are mounted directly onto a board without needing to be inserted into pre-drilled holes. This method is faster and more efficient than THT but requires a more precise soldering technique. It is also more difficult to modify the components once they have been mounted.

Through Hole Technology (THT)

This is the traditional method of soldering which simply involves inserting components onto a printed circuit board and then soldering them in place.This method of assembly is used for larger components such as connectors and switches. The components are inserted into holes that have been pre-drilled and then soldered with a soldering iron. Although this process is time-consuming and requires a high level of skill to ensure that all the components are correctly soldered and secured, it is relatively simple, requiring nothing more than patience and practice to master.

In this lab we will use THT because it is the simplest, most common and established method of circuit board assembly. It is also the:

- Most reliable since it involves soldering components directly to the board.

- Most cost effective since it involves fewer components and steps.

- Easiest to troubleshoot and repair, since THT board components are visible and easily accessible.

Objective

Use a wide, flat Chisel Tip and a Curved Bevel Tip to practice soldering components to a PCB.

Materials

In addition to your personal protective safety gear, gather the necessary tools and materials you will need to do this lab.

- Single Row Straight Header Strip Socket, 2.54 Mm Pin Heads

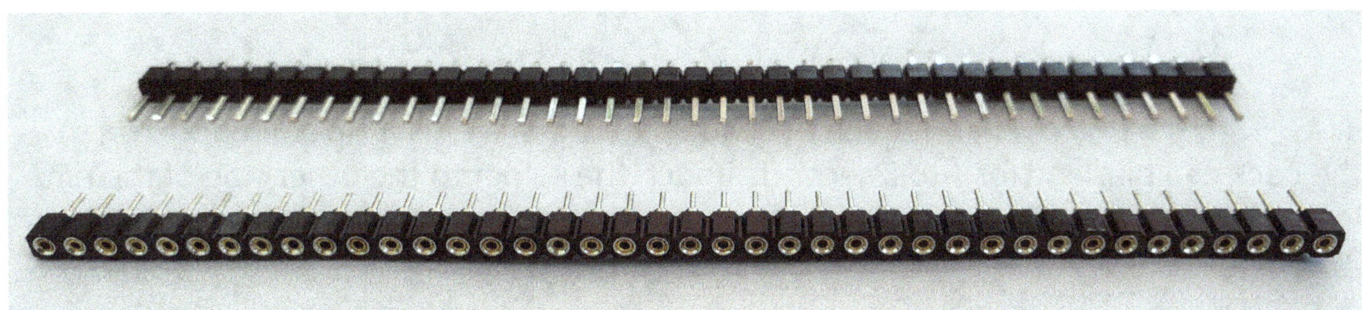

- Strip Sockets come in many sizes that will work, and you can even substitute other THT components for this Lab. We recommend these because they are inexpensive, readily available, versatile to any size board, and provide many pins for practicing.

- Practice PCB board with holes

- Acid-Free Rosin Liquid Flux and Small Paint Brush to apply Flux, or Flux paste

- Lead-Free Rosin-Core Solder

- Soldering Iron with Chisel Tip and Curved Bevel Tip

- Tweezers

- Needle-Nose Pliers

- Soldering Clips/Helping Hands or pliers to hold the wires (optional)

Instructions

1. Push the leads through the holes in the appropriate position on the PCB, and bend one or two leads with Tweezers or Needle-Nose Pliers, if necessary, so it is secure.

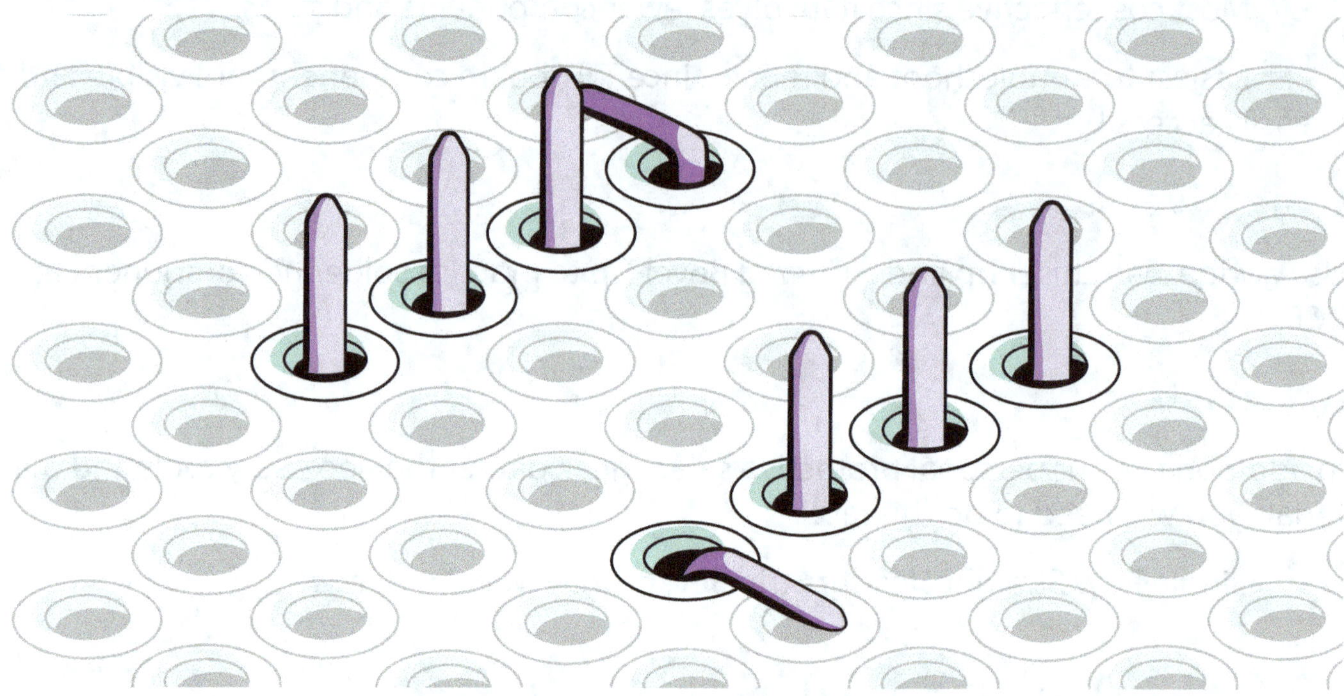

2. Turn the PCB over and touch the wide flat side of a Chisel Tip soldering iron to the back of a pin/lead for 1 to 3 seconds. Touch the solder to the other side of the pin/lead and allow the solder to melt and flow around the lead.

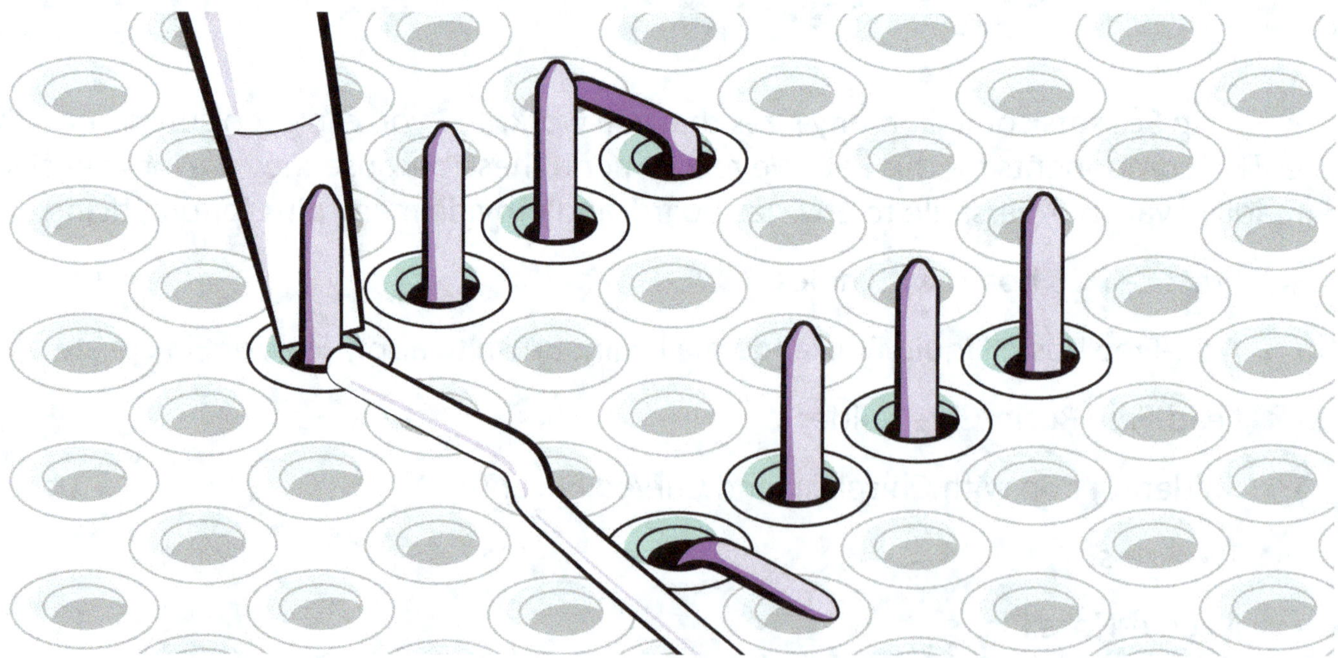

3.   Remove the solder and the iron. The solder should be in a mound shaped like a chocolate kiss, completely surrounding the pin or lead.

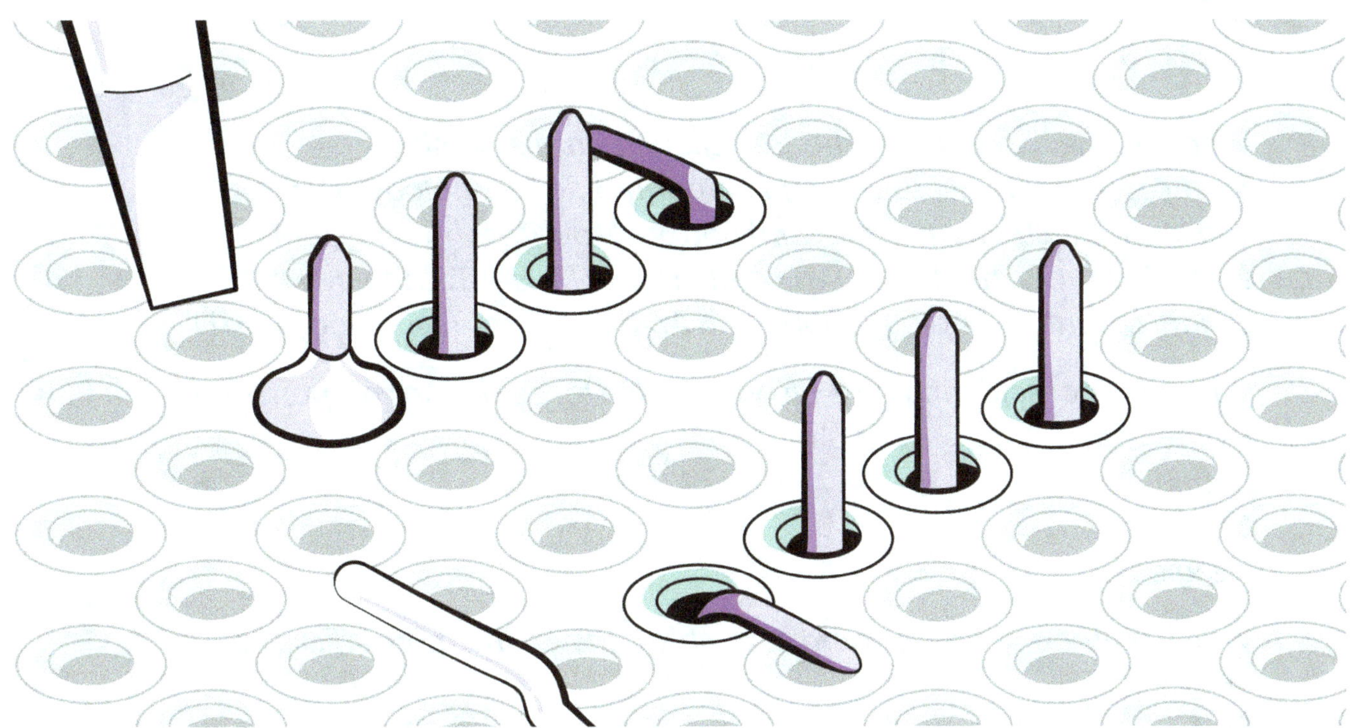

4.   Repeat this process until all the leads have been soldered.

5.   Repeat this process using a Bevel Tip also known as a Hoof tip to solder the sockets into place. Press the flat side of the bevel tip to the lead/wire. Decide which tip you prefer and write it here: _______________________________

6.   Finish the joint: After the joints of the component have cooled, use wire cutters to trim any excess solder, wires or leads.

7.   Once you are finished, turn off the soldering iron and clean the tip with a damp sponge or cloth.

Congratulations you have just soldered your first Through Hole Technology (THT) component! There are many inexpensive THT components and different styles of PCBs available to use if you desire more practice. After you've learned to solder the next step is learning about electrical circuits.

Check on Learning: THT

1.	What are the two main methods of circuit board assembly?
	a.	Surface Mount Device (SMD) and Through Hole Technology (THT)
	b.	Top Mount Technology (TMT) and Bottom Mount Technology (BMT)
	c.	Direct Mount Technology (DMT) and Indirect Mount Technology (IMT)
	d.	Digital Mount Technology (DMT) and Analog Mount Technology (AMT)

2.	Which type of soldering tip is recommended for soldering components onto a PCB?
	a.	Pointed tip
	b.	Round tip
	c.	Chisel tip
	d.	Needle Tip

3.	Which method of circuit board assembly is faster and more efficient?
	a.	THT
	b.	SMD
	c.	Both methods are equally fast and efficient
	d.	None of the above

4.	Which method of circuit board assembly is more difficult to modify the components once they have been mounted?
	a.	THT
	b.	SMD
	c.	Both methods are equally difficult to modify
	d.	None of the above

5.	Why is THT the most reliable method of circuit board assembly?
	a.	It is easier to modify components once they have been mounted
	b.	It is faster and more efficient
	c.	It requires a more precise soldering technique
	d.	It involves soldering components directly to the board

6.	Which method of circuit board assembly involves inserting components onto a printed circuit board and then soldering them in place?
	a.	SMD
	b.	THT
	c.	Both methods involve the same process
	d.	None of the above

Lab 6: Desoldering

In the world of electronics there are times when a component must be desoldered, so it's important to know how. For example, a faulty capacitor or a burnt out resistor could need to be desoldered and replaced with new components to fix an issue.

Additionally, during the build process, if two solder balls touch each other on a circuit board, it can lead to a short circuit, which can damage the circuit board and components. In such a case, you need to know how to clean up the board by removing any unwanted solder.

Objective

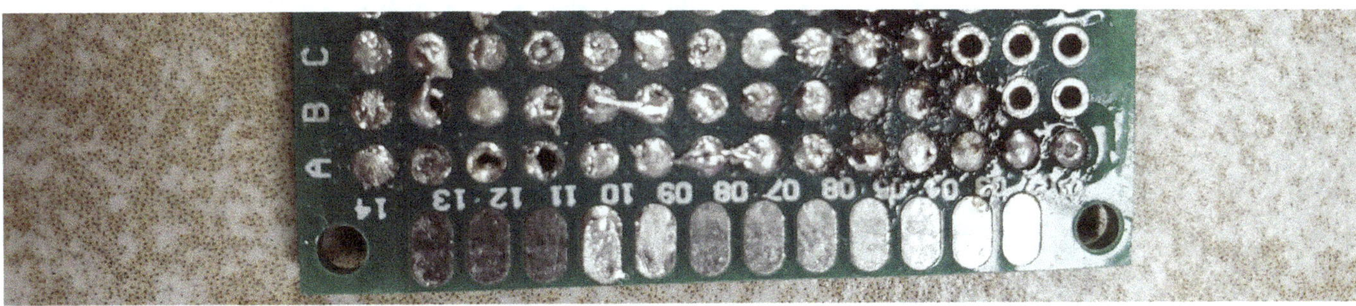

Let's say your drone is malfunctioning. You've done your trouble-shooting and have determined that a transistor needs to be replaced. Use a Chisel Tip or a Bevel Tip to desolder components. Practice using the Desoldering Wick and the Desoldering Pump to remove unwanted solder from a PCP board.

A.   Heat the solder with a soldering iron until it melts.

B.   Use a braided solder wick to move meted solder, or a desoldering pump to suck up the melted solder.

   a.   Press desoldering pump suction to arm.

   b.   Press the release button to create suction.

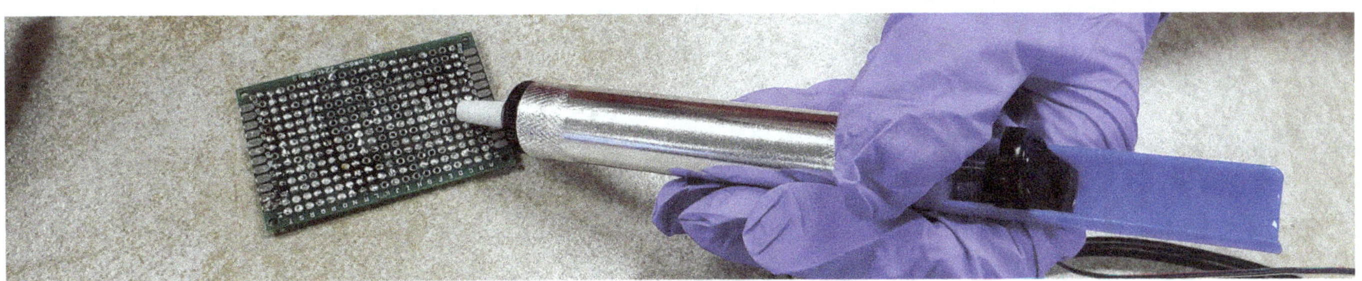

C.   If necessary, use a desoldering braid to further clean up the circuit board.

Materials

In addition to your personal protective safety gear, gather the necessary tools and materials you will need to do this lab.

- Practice PCB board with holes that has been soldered

- Soldering Iron with Chisel or Bevel Tip

- Desoldering Pump/Suction Tool

- Desoldering Wick/Braid

- Alchol/ Flux Remover, and cotton swabs

Instructions: Desoldering with a Suction Tool

A desoldering pump is a suction tool used in electronics repair to remove solder from a circuit board. It is a tube with a spring-loaded piston which gets pressed down and locked in place, until a button is pressed which releases the piston, thereby sucking up the air, and also the melted solder with it.

1.   Press the piston of the desoldering pump down.

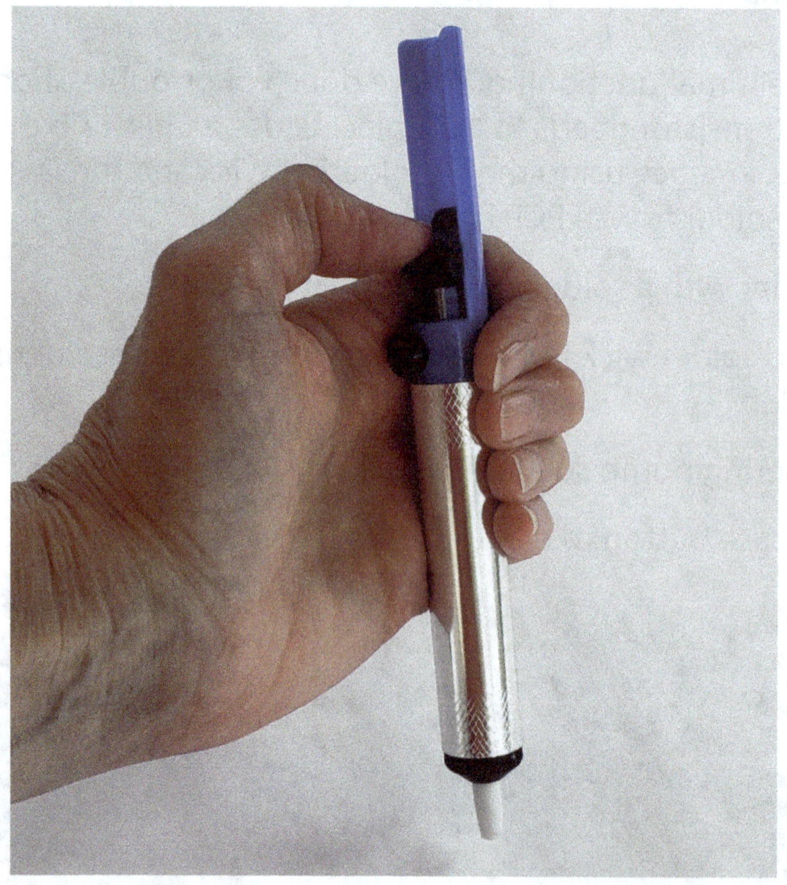

2.   Heat up the solder joint with a soldering iron until it melts.

3.   While the solder is still molten, position the desoldering pump over the solder joint with the nozzle tip touching the solder. Be careful not to melt the tip of the suction tool with the hot iron.

4.   Press the button or trigger on the desoldering pump to create a vacuum. The vacuum sucks the molten solder up into the pump.

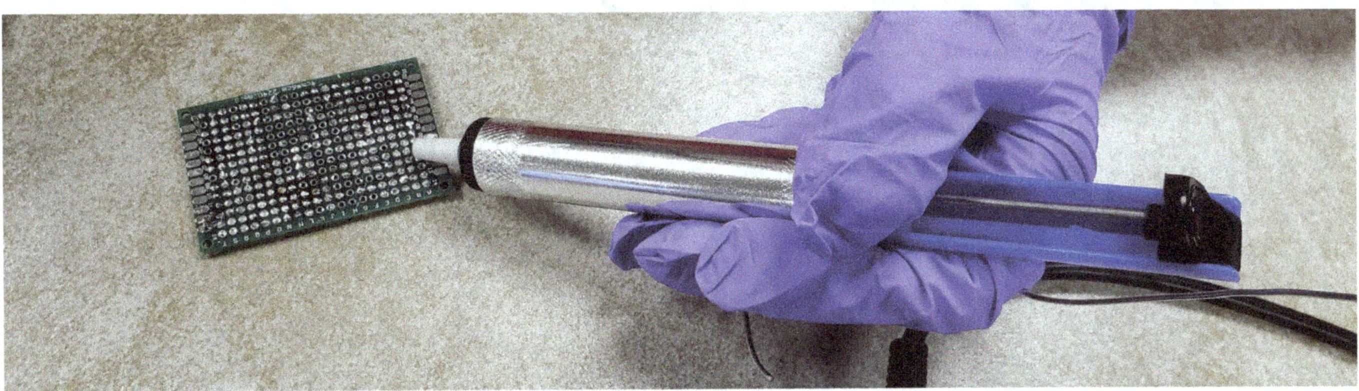

5.   Once all the solder is removed, clean the area with alcohol and a cotton swab or flux remover (available in liquid, pen or spray) and a clean brush.

Instructions: Desoldering with a Desoldering Wick

A desoldering wick, also known as solder wick or desoldering braid, is a tool used in electronics repair to remove solder from a circuit board. It is a flat, braided copper wire coated with a flux material that comes in different widths. It is used by heating the braid with a soldering iron, which melts the solder, and is then absorbed into the braid.

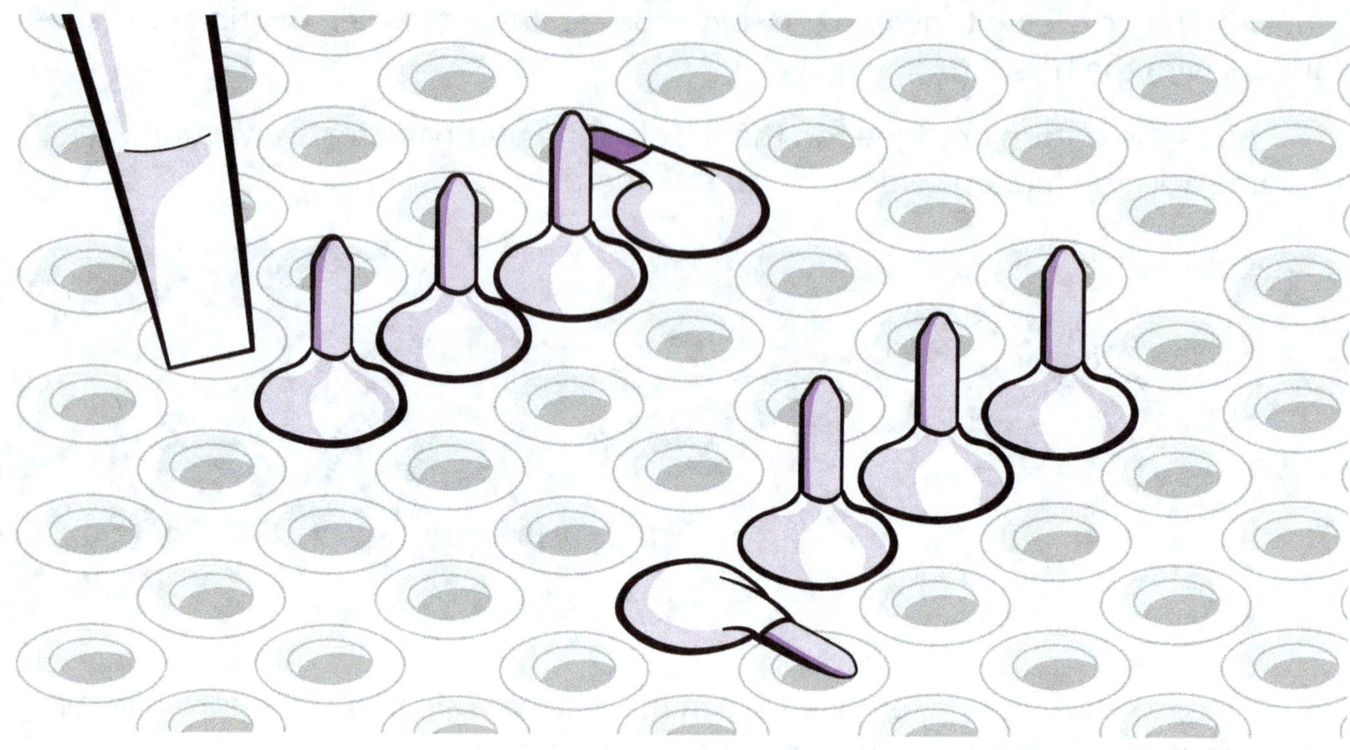

1.   Heat up the soldering iron and place the desoldering wick over the solder joint you wish to desolder.

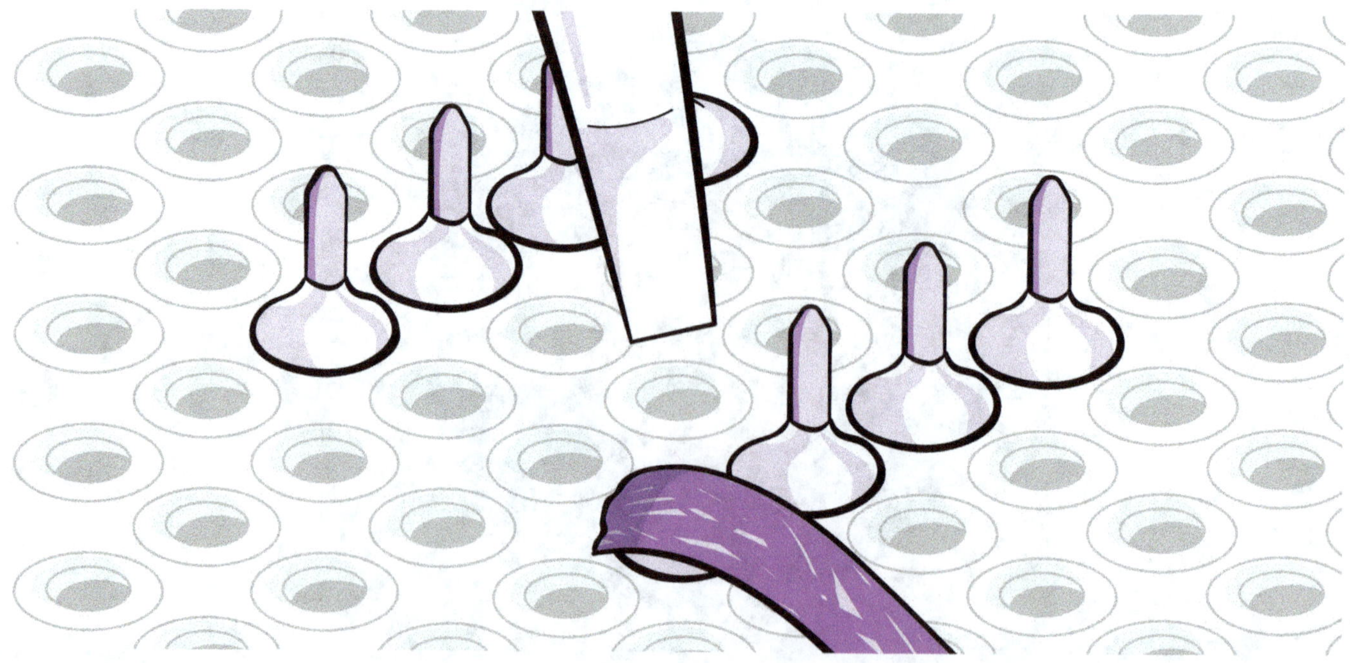

2.   Press the hot iron on the wick and keep contact with the desoldering wick until the solder is absorbed into the wick. The flux coating the wick should smoke and burn

away, as the wick becomes saturated with solder. If you have a larger area to desolder, press the flat side of the iron onto the wick.

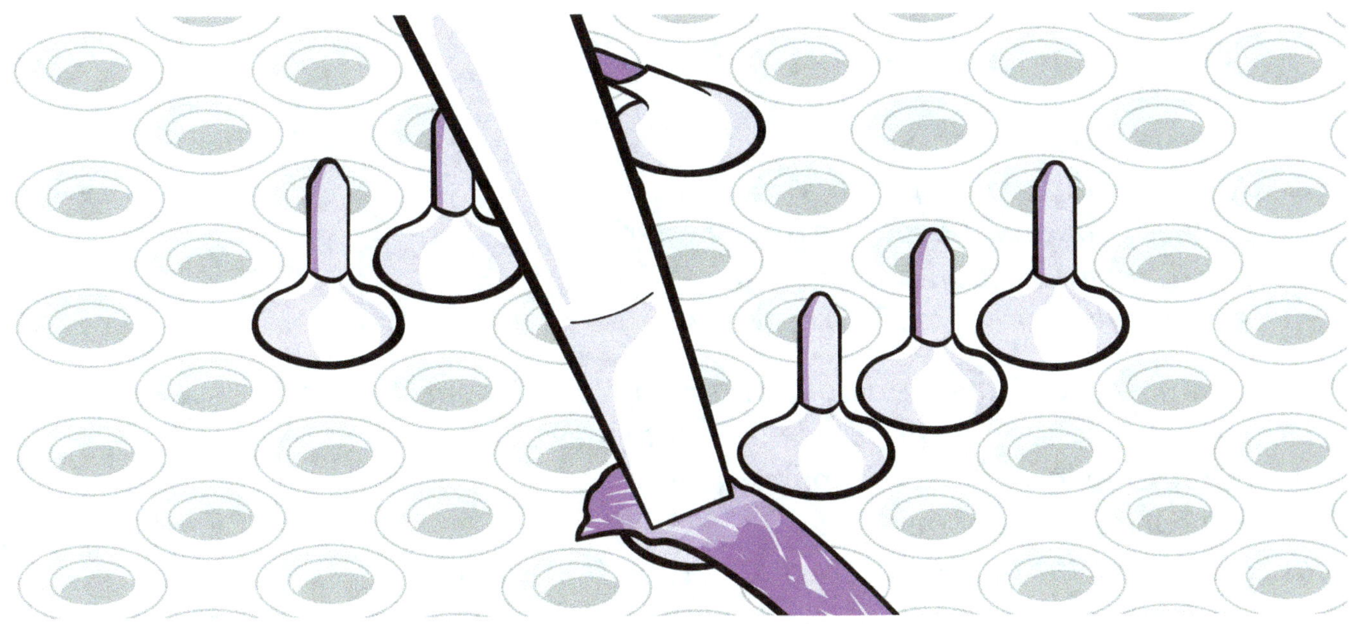

3.   Remove the iron and the desoldering wick, and inspect the area to make sure all the solder has been removed. If necessary, repeat Steps 1 and 2 until the desired amount of solder is removed.

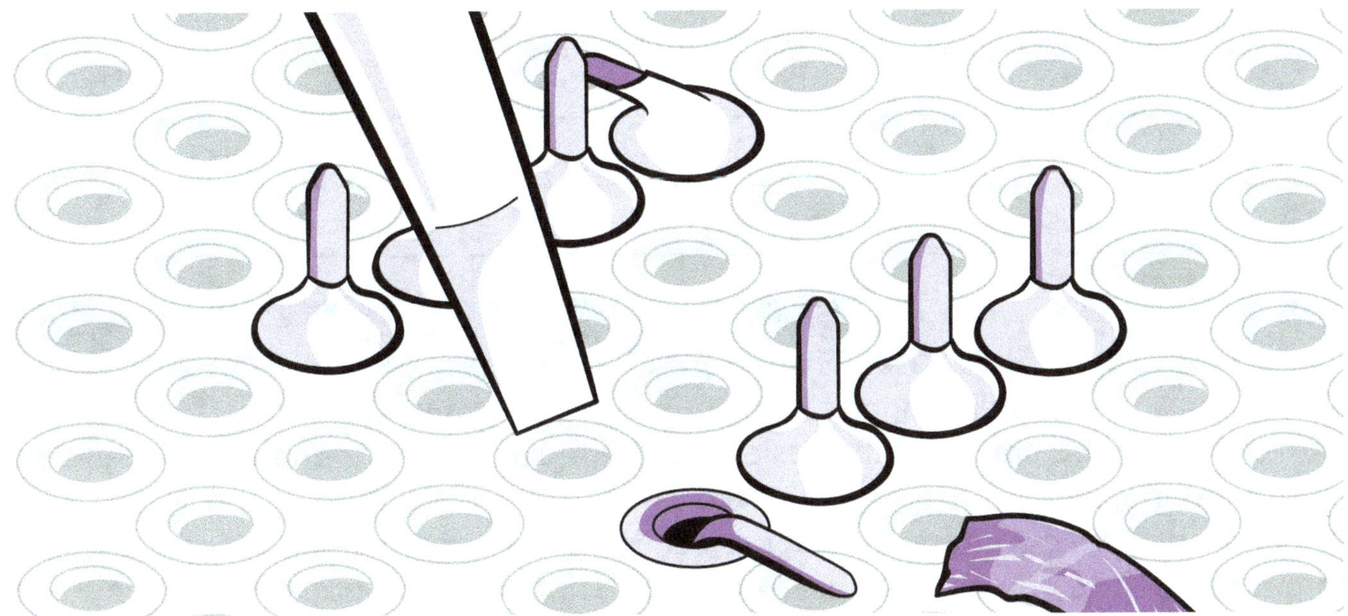

4.   If the solder joint is difficult to desolder, you can use a desoldering pump to help remove the solder.

5.   Once all the solder is removed, clean the area with alcohol and a cotton swab or flux remover (available in liquid, pen or spray) and a clean brush.

Check on Learning: Desoldering

1.   What is a desoldering wick?
   a.   A vacuum tube with a piston
   b.   Braided copper wire coated in a flux material
   c.   Plastic tubing filled with solder
   d.   A brush with bristles made of copper wire

2.   How do you activate the desoldering pump's sucking mechanism?
   a.   By pressing the tip of the pump down onto the PCB
   b.   By pressing the piston of the pump down
   c.   By heating up the pump
   d.   By pressing the trigger/button that releases the piston

3.   Why is it not good for two solder balls to touch each other on a circuit board?
   a.   You can't get the solder apart.
   b.   It looks messy.
   c.   It could create a short circuit which may damage components.
   d.   You waste valuable solder.

4.   Which Desoldering technique do you prefer and why?

____

____

____

Conclusion

Soldering can be an intimidating task if you don't know what you're doing. However, with this book, you now have the basics and can move forward with the confidence to start soldering. With practice and patience, soon you'll be a confident and competent solderer, able to tackle any soldering project. So, take some time to practice, and don't forget to wear the proper safety gear when soldering. With this book in hand, you're ready to start your soldering journey.

Next steps

As a next step, you can look into more advanced soldering techniques, such as adding components to your circuit board, soldering with multiple components or working with surface mount components. It's also time to start learning about circuits, so you can understand the how and why concerning the flow of electricity through the components you solder. Start with something easy and practice, practice, practice. Good luck and happy soldering!

1.   *What is the best way to become good at soldering electronics?*
   a.   *Try to build a computer.*
   b.   *Do lots of different types of soldering with a variety of equipment.*
   c.   *Work fast to get lots of experience.*
   d.   *Take your time, have patience, and practice.*

2.   *List the personal protective equipment (PPE) recommended by this book.*

____

____

____

3.   *Summarize what you have learned about soldering throughout this book and how you can apply these techniques in the future?*

____

____

____

____

____

4.   *What soldering projects are you eager to complete?*

____

____

____

____

____

____

Crossword Puzzle

Across

2. Worn for safety

4. Don't touch until it

6. Hold tool like a

8. Add heat until it

9. Caused if two balls touch

11. Toxic substance

12. After using iron

13. Most popular tip

Down

1. Shape of solder mound

2. Put on hands

3. Abbreviation for tin

5. Cleans the tip

7. Heat to 675 degrees

10. Coating tip of iron with solder

```
        I  L  P  S  P  O  O  L
      I  J  P  D  G  C  O  N  I  C  A  L
    Y  C  L  O  J  U  O  Y  P  U  M  P  H  R
  N  U  D  J  E  I  B  O  A  R  D  Y  W  R  A  L
  T  G  T  A  E  E  A  N  U  W  R  N  F  N  P  L  F  O
  S  H  O  U  R  S  J  D  T  L  H  I  S  V  B  A  L  L
W  O  R  X  A  T  W  O  P  R  I  S  O  L  D  E  R  I  N  G
F  L  Y  I  O  T  C  M  L  Z  S  L  E  E  V  E  S  C  H  K
C  D  D  D  R  I  S  K  U  D  Q  I  Z  R  M  O  L  T  E  N
G  E  S  I  T  N  B  T  I  N  E  J  O  R  O  S  I  N  K  C
C  R  P  Z  A  N  H  K  G  R  W  R  N  Z  Q  M  P  L  D  H
D  Q  O  E  K  I  T  E  L  M  O  P  I  C  I  R  C  U  I  T
Q  C  N  S  D  N  W  W  O  N  E  N  Y  N  L  S  C  L  A  Q
U  P  G  C  E  G  E  B  V  F  L  T  U  S  G  A  D  Q  I  D
  K  E  H  G  B  E  V  E  L  S  B  A  E  O  F  S  A  F
  L  U  A  R  B  Z  C  S  T  J  K  R  L  N  E  O  C  N
    O  Z  E  O  E  X  I  I  B  C  M  A  A  T  D  B
      D  E  N  R  G  L  A  S  S  E  S  I  Y  S
        S  D  S  H  E  A  T  J  L  F  X  D
          L  P  H  I  H  E  T  R
```

BALL	CONICAL	IRON	OXIDIZE	SOLDERING
BEVEL	DEGREES	JOINT	PUMP	SPONGE
BOARD	DESOLDERING	LEAD	ROSIN	SPOOL
BOND	GLASSES	MELT	SAFETY	TIN
BRAID	GLOVES	METAL	SLEEVES	TINNING
CIRCUIT	HEAT	MOLTEN	SOLDER	TWEEZERS

Answer Keys

Answer key for "Solder and Flux "

5.	d

6.	b

7.	a

8.	b

Answer key for "Soldering Tips "

1.	d

2.	a

3.	Bevel Tip, Knife Tip, Conical Tip, Chisel Tip, Needle Tip

Answer key for "Safety "

1.

Gloves

Safety glasses

Long sleeves

Tie back long hair

Heat resistant holder

First aid kit

Fire extinguisher

2.

Inhalation

Skin contact

Ingestion

3.

Stone

Ceramic Tile

Stainless Steel

Silicon

Answer Key for "Soldering"

1.	a

2.	a

3.   c
4.   b

Answer Key for "Tinning"

1.   b
2.   c
3.   b
4.   b

Answer Key for "PCB Board "

1.   c
2.   a
3.   a
4.   b

Answer key for "Soldering Wires Together "

1.   b
2.   c
3.   a

Answer Key for "Surface Mount "

1.   b
2.   False
3.   A. Strip and/or Twist; B. Flux; C. Tin

Answer Key for "THT "

1.   a
2.   c
3.   b
4.   b
5.   d
6.   b

Answer Key " Desoldering"

1.   b
2.   d
3.   c
4.   Desoldering Wick or Desoldering Pump

Answer Key for "Next Steps"

1.   d
2.   Safety Glasses, Gloves, Long Sleeves, Hair Band